Sandyooo

With best wishes

E Mott Maynard

Transforming the Global Biosphere

Twelve Futuristic Strategies

Revised Edition

By Elliott Maynard, Ph.D., CPCM

Published by

Arcos Cielos Research Center
Sedona, Arizona USA
www.arcoscielos.com

Transforming the Global Biosphere
Twelve Futuristic Strategies
Revised Edition by Elliott Maynard, Ph.D.

Published by
Arcos Cielos Research Center
Postal Drawer 20069
Sedona, Arizona 20069 USA
arcoscielos@yahoo.com
www.arcoscielos.com

ISBN 13: 978-0-9721713-2-8
ISBN 10: 0-9721713-2-0

Printed as an Arcos Cielos hardcover
by arrangement with Watchmaker Publishing, Ltd.

Printed as an Arcos Cielos softcover by arrangement with Personal JAZ.

Interior re-design by Donald Enevoldsen

Cover re-design by Larry Walker, Image Legends

Production coordinated by Janet Jaz

Dedication

This book is dedicated to the new breed of global leaders and luminaries everywhere. It is my hope and vision that these Change Masters will unite in a new spirit of dedication and collaboration to empower new programs and technologies for a positive transformation of Earth's biospherical and social consciousness. I like to call this noble odyssey we have embarked on, "The Magnificent Revolution."

I express my deepest gratitude to my former wife and mission partner, Sharon Maynard, for all the magical gifts, inspiration and loving support she has bestowed.

CONTENTS

ACKNOWLEDGEMENTS

The Author wishes to acknowledge the following individuals, whose kindness and support have made this book a reality: Todd Barber, CEO of the Reef Ball Development Group, Inc.; Lt. Col. Thomas E. Bearden, Ph.D., (U.S. Army Ret.), Roar Bjerknes, Founder and Chairman of R. B. Media Pte. Ltd. (The *Gaiaship* Project); Kenneth J. Cox, Ph.D., Chief Technologist for HEDS at NASA's Johnson Space Center; Mike Crosby, Executive Vice President of SkyCat Technologies, Inc.; Roger Davey, Executive Director of Energen Global, Inc.; Jacque Fresco and Roxanne Meadows, Founders of the Venus Project; Constantin I. Ivanenko, Ph.D. Founder of Russian Initiative Group for Defense of Earth; Andrew Michrowski, Ph.D. President of the Planetary Association for Clean Energy; Robert A. Nelson, President of Rex Research; Ruggerio Maria Santilli, Ph.D., President of the Institute for Basic Research; Dennis Stallings, President of Aerobus International; Vito Tanzi, CPCM, Founder and Chairman of the National Bureau or Certified Consultants; Thomas P. Valone, Ph.D., PE, Founder and President of the Integrity Research Institute; and Lt. Col. Edward E. Winchester, CM. (U.S. Army Retired), Founder and Past President of the Pentagon Meditation Club.

My heartfelt thanks also, to Selma Brackman, Executive Director of the War and Peace Foundation, who suggested I send my original World Future Society Presentation to former UN Assistant Secretary General, Dr. Robert Muller. Finally, I would like to express my deepest appreciation to Dr. Muller himself, for his kindness, wisdom and mentorship in encouraging me to write this book and especially, for enriching my life by sharing his "Dreams and Ideas for a Better World."

FOREWORD

Transforming the Global Biosphere is a holistic, extremely insightful analysis of the urgent problems we face today. This book is a call to action to re-order our priorities. It offers a workable and attainable blueprint for society that meets the needs of all people, while protecting and preserving the Environment.

Elliott Maynard makes it incontestably clear that criticism of our culture's insufficiencies is not enough, unless alternative methods are presented to address these problems. This excellent book thus serves as a guideline for the protection of the environment, and for creating a sustainable future for everyone.

Transforming the Global Biosphere is not only a major contribution to the field of Future Studies; it also deserves to be read by all individuals who are concerned with the contemporary cultural climate. A significant achievement of this work lies in its wealth of information in areas such as clean and renewable sources of energy, adequate transportation, new and innovative approaches to food production, restoration of the environment, and many other essential elements for a positive future scenario.

This book is without a doubt one of the most relevant contributions presently available for helping to solve our current social and ecological problems. Elliott Maynard is among the few individuals with the depth and breadth of vision that will enable us to achieve a sane and sustainable future. His book is written with a deep sensitivity and concern for the well being of all people everywhere. The subject matter goes directly to the core of our global crises, and presents practical alternatives for a sustainable lifestyle for everyone. Essentially, it represents a ground breaking vision for a more humane future—a Future derived from the advantages of using

Science and Technology intelligently. Everyone will be enriched by reading this hope-filled, highly recommended treatise.

Transforming the Global Biosphere should serve to stir society into seriously addressing and confronting the devastation that is presently impacting our environment. Maynard makes it incontestably clear that we can restore the land, sea and air to as near a natural condition as possible, while providing clean sources of renewable energy.

Elliott Maynard's straightforward and well-documented description of the seemingly insurmountable problems of today has taken a wide vision approach. Through years of study, he has accumulated a significant amount of relevant information, from which he creates a series of positive alternatives for avoiding further environmental degradation and unnecessary human suffering. This is a groundbreaking, highly recommended treatise that presents a workable blueprint for a bright and more humane future for everyone.

~ Jacque Fresco, Futurist, Architect, and Director of *The Venus Project,* Venus, Florida, 2003.

Note: Jacque Fresco is a forerunner in the fields of Industrial Design and Human Factors Engineering. Throughout his professional career he has developed a vast array of creative and innovative designs which include prefabricated houses, automobiles, electronic and medical equipment, and hundreds of commercial products and inventions. He has also served as Technical Advisor for several motion pictures, and is the author of many futuristic books and videos. His concepts have been presented on numerous radio and talk shows, and he has also been featured in many national and international magazines and newspapers.

INTRODUCTION TO THE SECOND EDITION

During the interval between the first edition (2003) and this second edition, many historically unprecedented shifts have occurred in the global arena. The original intent of this book was to put forth a set of basic tools and strategies as unique new paradigms for social and ecological transformation, and to suggest creative ways to ground these new paradigms into practical reality. Today, this information is even more relevant than ever for the continued success and survival of the human race.

We continue to experience a cascade of major shifts in our social, financial and ecological structures. The good news is that this turbulence has sparked a new breed of global leaders and luminaries. They quietly come forth and unite within a new spirit of enlightened collaboration in order to bring new paradigms for positive change into practical reality. For those brave individuals who are willing to work together synergistically, these are indeed times of great challenge—but also times of great opportunity.

The global economic crisis has underscored the critical importance for weaning society from its addiction to polluting and environmentally destructive fossil fuels; and for working to discover new sustainable ways to bring alternative energy and other advanced technologies into mainstream society. As a species, we can no longer continue to exist within the ecologically dysfunctional paradigms of the past. Instead, we must work together to create new open-ended paradigms, which are flexible enough to evolve and adjust according to the needs of any particular time-point.

From a human consciousness perspective, we must shift from our outdated linear, rear-view-mirror-thinking modality, into a new realm of quantum-field thinking. I am deeply convinced that such a shift will open bright new pathways into the future. Such a profound shift in our consciousness is critical for establishing an entirely new and ecologically appropriate mindset for the health and sustainability of our Global Biosphere, as well as for the happiness and well-being of all humans on Planet Earth. The very survival of our species depends upon it, as the legacies we create today will become the precious gifts we pass on to the generations of tomorrow.

Some of the technologies introduced in the first edition have gone through changes and corporate restructuring, while others have disappeared entirely—being replaced with newer and more appropriate concepts. The original concepts and ideas in the first edition can be best updated by entering each concept or technology into an internet search engine to discover their current status.

This second edition includes most of the original websites and references in the first edition, but new links have been added in parenthesis to update the information. Several key books and DVD's have also been added to the Bibliography to highlight new advanced technologies.

For readers wishing to learn about the latest advances in Alternative Energy Technologies, (Metastrategy IV), I would suggest the following three books: *The Energy Solution Revolution* by Dr. Brian O'Leary (2008), *Break Through Power: How Quantum-Leap New Energy Inventions can Transform our World* by Jeane Manning and J. Garbon (2009), and *Zero Point Energy: The Fuel of the Future*, by Dr. Tom Valone (2009). These books complement each other nicely and provide a comprehensive overview of the possibilities for leapfrogging directly into these advanced energy technologies—especially in areas of zero-point energy and other leading-edge areas of science.

Within this advanced realm of scientific thinking, I would agree with Dr. Brian O'Leary, who suggests that "renewable energy technologies" such as solar, wind power, wave power, tidal power, and biofuels will, in the future, come to be regarded more as transitional measures, when the overall environmental costs are factored in. What is needed is a new sustained global movement to bring advanced zero-point energy and other advanced technologies into mainstream applications everywhere. This would provide non-polluting, cost-effective, and environmentally sustainable solutions for electrical power generation, transportation, and waste and water recycling.

This said, in light of recent information obtained by the author at the Third International Conference on Future-Energy (Washington, DC, October 9-10, 2009), the newest wild-card science will involve the integration of innovative new applications of water-fuel technologies, which will facilitate the integration of "charged water-gas clusters" into existing internal combustion engines. Functional laboratory and commercial units, which involve Electrolyzer and Brown's Gas Technologies, have already produced demonstrable and patentable results. They appear to be tapping into the Quantum Vacuum, as they demonstrate over-unity energy production in ways which defy the laws of conventional science. An overview of these new technologies can be found by entering "Water Fuel ZPE" into your Internet search engine, or at: http://peswiki.com/index.php/Video:Water_as_Fuel_(via_ZPE).

The time has come for each of us to stand up and do the best we can each day, to make this world a better place. In facing up to the present global challenges, we must effectively combine global thinking with individual participation, and work together to birth a new "Magnificent Revolution." To accomplish this, we must first break the shackles of the established mindsets of greed, scarcity, competition, pollution of the global commons, and the wanton squandering of Earth's natural resources; and shift into new enlightened scenarios of cooperation, sustainability, abundance, and social justice for everyone on Earth. This is our "critical protocol" for survival and success!

This is the time for decisive action, and each of us has the opportunity to become key players in this new and exciting quest for global transformation. "Miracles" are most often the result of countless tiny steps, which eventually reach a "critical mass" to power a major positive shift in global consciousness. In this way, each of us can become "co-creative miracle workers" by simply taking one or more tiny positive steps ahead each day. Over time, and amplified by millions of individuals taking action everywhere, this kind of transformation will not only be possible; it will be inevitable.

~Elliott Maynard, 2009

INTRODUCTION

The Technosphere vs. the Biosphere

**If an environmental crisis is the trigger needed
to detonate
an Evolutionary Time Bomb,
then our generation
certainly qualifies
for a Transformation of the Species.**

~ Yatri, *Unknown Man: The Mysterious Birth of a New Species.*

Global Ecology is a subject of such overwhelming magnitude and complexity, that it is truly difficult to fully comprehend its dynamic fabric of nuances and interactions. Despite our relatively advanced level of technical sophistication, we have made only minimal progress in reversing the destructive environmental trends caused by the impacts of human activities on the Global Biosphere. A transformative revolution on a global scale requires an entirely new set of guidelines for thinking and living. Such a revolution must incorporate sweeping social transformations, where every inhabitant of planet Earth participates, by living within an inspired and self-imposed "Global Constitution of Sustainable Limits"—a set of guidelines, which is environmentally synergistic with our global ecosystems. If we are truly committed to achieving this transformation, and to blazing bold evolutionary pathways for humanity into the future, a new and environmentally focused planetary mindset must be established and implemented at all levels of human society.

Even in ancient times, human activities and technologies were responsible for major changes in the global ecosystems. Twelve thousand years ago, migrations of Siberian hunters into the Americas coincided with the rapid waves of extinction of

21

80 percent of the bird and mammal species that had thrived since the Late Pleistocene epoch. Everywhere early man established himself, similar mass extinctions occurred.[1] According to Global Ecologist David Suzuki, recent scientific evidence indicates that human activities *were directly responsible* for these mass extinctions—and for similar mass extinction that occurred in Australia some 35,000 years ago.[2]

Six to eight thousand years ago, the great Mesopotamian Empire was established. In the 1500 years which followed, however, 90 percent of the lush forests of the surrounding regions were destroyed by human activities. This type of massive ecological devastation was largely responsible for the eventual downfall of this mighty civilization. Similar massive ecological disruptions resulted in the eventual collapse of both the Greek and Roman Empires.[3]

The major environmental problems which presently face global society relate directly back to our Industrial-Age mindset and our ecologically inappropriate technologies. Historically, humans have regarded our planetary resources as inexhaustible assets—existing to be harvested and controlled by the most powerful individuals and institutions of the times. If humanity is to survive, and continue evolving successfully throughout the 21st Century (and beyond), its current social overlay of environmental dysfunctionality must not be allowed to continue in its present form. Our presently self-centered, consumption-driven, and wasteful Industrial-Age mindset needs to evolve into a new transformative and self-generating life-ethic—a worldview which highlights cooperation, social justice, and spiritual fulfillment. A new "Tri-Millennial Mindset" would provide a fresh emphasis on creativity, cooperation, and environmental sensitivity, with the responsibility being shared by individuals, institutions, and nations alike.

In order to reverse the presently destructive impacts of human technology (the technosphere) on the Global Biosphere, the author proposes Twelve Futuristic MetaStrategies for implementation on a global scale. Each of the following MetaStrategies represents a key Global SocioEcological paradigm in itself. Within each paradigm are cited examples of major problems and abuses to the planetary environment. Innovative concepts and unique new alternative technologies are also presented for reversing these destructive ecological trends. Many of these new paradigms combine existing technologies with new sustainable technologies. Although each Metastrategy represents a specific "Perspective of Interactions" between current socio-technological behavior, and the Global Biosphere, the MetaStrategies overlap and intertwine to form a "Dynamic, Interactive Fabric." When taken as a whole the Twelve MetaStrategies constitute a set of conceptual tools for creative thinking and action, designed to promote enlightened and environmentally sustainable futures; future global scenarios, which universally benefit humanity, its planetary homeworld, and the generations yet to come.

> **Wild dark times are rumbling toward us,**
> **and the prophet who wishes**
> **to write a new apocalypse**
> **will have to invent entirely new beasts,**
> **and beasts so terrible**
> **that the ancient symbols of St. John**
> **will seem like cooing doves and cupids in comparison.**
>
> ~ Heinrich Heine, 1797-1856

Earth at a Critical Environmental Crossroads

The United Nations Environmental Programme (UNEP) recently declared that global society had reached an "Environmental Crossroads," where a choice between self-serving greed and a positive future for humanity will decide the fate for millions of people in the decades to come.

In the words of UNEP Executive Director, Klaus Toepfer, "Fundamental changes are possible and required." The recent GEO report, which preceded the 2002 Earth Summit Meetings in Johannesburg, South Africa, paints a dismal future for the global

environment, and for one single species—**H*omo sapiens!*** At
this time, there seems to be an increasing awareness among the
cognoscenti of global society, that if lasting solutions to our current "Ecotastrophe" are to be discovered and resolved, the causes
and cures for this planetary transformation must emerge from deep
within the collective hearts and minds of humans themselves.[4]

Environmental Issues and Global Security

Lloyd Axworthy, Chairman of the Manitoba Task Force on
Climate Change, has pointed out the direct link between the environment and global security. In the war against terrorism, a
major effort to reduce North America's dependency on fossil fuels
would seem to be a good strategy, since a primary objective of
fundamentalist Islamics is to get Americans to leave the Persian
Gulf Region.

Thus, if North Americans could improve both their Energy
Efficiency and Conservation—while concurrently developing new
Alternative Energy Resources—we would become much less dependent on oil from the Middle East, and thus remove a primary
motive for terrorist threats.[5]

Global Warning!! Everest, Kilimanjaro, and Peruvian
Glaciers Melting

A recent UNEP scientific expedition to Mt. Everest discovered that the glacier where Sir Edmund Hillary set out for his
historical climb nearly 50 years ago has retreated up the mountain
slope nearly three miles! The research team of mountain climbers
reported that the impact of rising temperatures was in evidence
everywhere; with the mountain terrain bearing the unmistakable
scars of sudden glacial retreat, and glacial lakes being swollen by
melting ice. During their trek, the team spoke with the directors of
the Nepal Mountaineering Association, who told them they had observed many rapid changes over the past 20 years. They indicated
that Hillary and Tenzing would now have to climb for another two

hours or more to reach the edge of the glacier that was formerly close to their original base camp. Recently, the UNEP issued an environmental warning that more than 40 Himalayan glacial lakes were dangerously close to bursting out of their confines, thus endangering the lives of thousands of people, due to the ice melt caused by global warming.[6]

According to Ohio State University Scientist, Lonnie Thompson, recent aerial studies of 19,000-foot Mt. Kilimanjaro in Tanzania, indicate it has lost approximately one-third of its ice fields over the past two decades. At the present rate, it is estimated that the rest of its ice cap could disappear entirely by 2015. Professor Thompson explained, "Glaciers are like natural dams. They store the snow in the wet season, and they melt in the dry season to bring water flow to the rivers."[7]

Recent scientific surveys of the Quelcaya Ice Cap in the Peruvian Andes indicate it has shrunk by 20 percent since 1963. More importantly, the retreat of its largest glacier has *accelerated* by 509 feet per year.[8] From a global ecology perspective, glaciers might be thought of as planetary-sized "Canaries in the Mines." As such, they function like massive computers, in that their condition at any given time integrates all existing climatological parameters, thus providing us with a basic picture of the general health of Planet Earth.

Global Warming Shatters Giant Antarctic Ice Shelf

In March of 2002, satellite photos revealed that a gigantic ice shelf the size of a small country had broken away from the continent of Antarctica. Although four years previously, scientists from the British Antarctic Survey (BAS) had predicted the eventual disintegration of the huge Larsen Ice Shelf B (some 1,255 square miles in area), they were astonished at the speed at which the breakup had actually occurred.

BAS Glaciologist, Dr. Vaughn expressed the reaction to this major ecological event as follows: "We knew what was left would collapse eventually, but the speed of it is staggering. It is hard to believe that 500 million billion tons of ice sheet has disintegrated in less than a month."

The Larsen B Ice Sheet was one of five Major Ice Shelves—huge masses of ice that are floating extensions of the Antarctic Land Mass. These ice shelves have been steadily shrinking in recent years, due to the effects of rising atmospheric and oceanic temperatures. Over the past 50 years, the Antarctic Peninsula itself has warmed by an average of 2°C—a far greater increase than elsewhere in Antarctica, or other parts of the world. Because the ice shelf was already floating, the breakup will fortunately not create any appreciable rise in sea level.[9]

**What is unprecedented about the present situation
is that it is the actions and technical productions
of one species—the Human Being—
that is bringing about this biosphere meltdown.
Increasing numbers of people
have therefore come to the conclusion
that it is in the hearts and minds of human beings
that the causes and cures of the ecotastrophe
are to be found.**

~ Ralph Metzner—1999, *Green Psychology*, p. 1.

Atmospheric Pollution:
The "Dark Force" vs. the Global Biosphere

The loss of sunlight from severe air pollution is apparently lowering temperatures over vast areas of the global biosphere, and threatening water supplies for millions of people. A massive cloud of "Brown Haze," mainly from coal-fired industry and biomass fires, has been blamed for blocking out up to 15 percent of the sunlight in Southern Asia.

Recent scientific studies have begun to paint a clearer picture of how atmospheric pollution affects the amount of sunlight falling on Earth's surface. What *is* apparent is that airborne pollution from fossil fuels and burning vegetation has been plunging vast areas of our planet into a chilly darkness—negatively impacting both agricultural crop yields and water supplies. Studies in China, for example, have shown that vast regions of the country have experienced a steady decline in sunlight, combined with steadily dropping average temperatures. In Brazil's Amazon Basin and in Zambia, during certain times of the year, air pollution is so bad it obscures nearly twenty percent of the sunlight reaching the Earth's surface.

Scientists Dale Kiser and Yun Qian, of the Pacific Northwest National Laboratory in Richmond, Washington, have been studying the effects of Coal Burning in China over the past 50 years, to determine the impact pollution-related aerosols have had on the amounts of sunlight reaching the ground. Since the 1960's China's sulfur dioxide emissions (mainly from coal-fired industries) have risen five-fold—with the result that China now emits as much sulfur dioxide as the United States. Using solar-powered monitoring devices to record the amounts of sunlight reaching the ground in different locations, the scientific team discovered there has been a significant decline in the amounts of sunlight reaching most areas of the country, with the pattern of decline matching the increased rate of fossil fuel burning over time.

According to Qian, pollution-induced aerosols do more than just block the sunlight. They also "affect human health and acidify the land." Reduction of solar radiation also negatively impacts agriculture, since reduced solar energy tends reduce agricultural crop yields. Reduced sunlight also means less evaporation, which creates less rainfall for crops and for replenishing basic regional water supplies.

In addition to the burning of coal and other dirty fossil fuels, soot and ash from grass and forest fires (started by farmers to clear the land, especially in the Amazon Basin), have contributed dramatically to reducing the amounts of sunlight reaching the ground by as much as 16 percent. The smoke from burning fossil fuels and biomass contains aerosols of sulfur dioxide, ash, and black soot particles that act in a variety of ways to reduce ground temperatures. The aerosol particles function as nuclei for forming water droplets. These droplets then form clouds, which reflect both sunlight and heat back into space. The ash and soot particles themselves can also shade the ground, while suspended soot particles have a tendency to absorb heat, creating unpredictable atmospheric heating effects, which in turn influence climate.

Although the cooling effects of atmospheric pollution over areas like China, the Amazon, and Zambia tend to counteract global warming to a certain extent, the overall negative effects on human health, agriculture production, and water resources provide additional reasons to cut back on greenhouse gas emissions. Aerosols tend to remain in the global atmospheric envelope for weeks at a time, while carbon dioxide levels can remain relatively stable for years.[10]

Many of Earth's Lakes Face Ecological Death

According to Will Cosgrove, President of the World Water Council, many of Earth's freshwater lakes face ecological death from pollution—an event which could have disastrous consequences for the humans that depend on them. Says Cosgrove, "We're killing the lakes, and that could be disaster to the human communities that depend on them." The majority of threats to these lakes directly relate to global population growth, plus the effects of development, industrialization, and pollution. Cosgrove also pointed out that one particularly insidious aspect of the ecological threats to the world's lakes is that, even though a lake may *appear* pristine, many lakes have already suffered long-term ecological damage from acid rain and nutrient enrichment from agricultural fertilizer runoff. According to Cosgrove, "Suddenly something happens like a change in water temperature, and all of a sudden a lake can be transformed. Once the process starts, it's hard to stop."

Said Cosgrove, "Humans are already using more than 50 percent of the usable freshwater resources, and 90 percent is from freshwater lakes." Despite the magnitude of the threat to this critical planetary resource, dealing with pollution and other ecological problems in freshwater lakes generally remains low on the list of governmental priorities.

Extreme examples of environmentally endangered lakes include Lake Victoria, Africa's largest lake, which over the past twenty years has witnessed the disappearance of several fish species, plus dramatic increases in aquatic plant growth, due to nutrient loading from different sources—including raw sewage from nearby villages. The resulting weed growth has been so prolific that local fishermen can no longer even get their boats out from shore into open water. Another extreme example of a seriously endangered lake is in Lake Taihu in China, said to be so heavily polluted, "that you can practically walk on it." Environmental deterioration of this magnitude tends to have major negative impacts on the regions surrounding large lakes, and directly affects the livelihood of nearby populations, resulting in poor nutrition, starvation, and illnesses which can be traced directly to contaminated drinking water.

One of the primary objectives of the World Water Council is to create a new heightened global awareness that the world's lakes are precious ecological resources, and that their delicate ecological balance must be protected for the well-being of present and future generations. It is hoped that this new ecological awareness, plus new environmental education programs, can create public opinion pressures that will stimulate effective action by governments around the world.[11]

Collapse of Northern European Cod Fishery Predicted

According to recent scientific reports from the International Council for the Exploration of the Seas (ICES), codfish populations from the waters off western Norway to the Atlantic shores of Scotland have become so depleted it was recommend that, "All fisheries in this area that target cod should be closed." Recent scientific reports have indicated that for the past 17 years, the numbers of codfish reproducing in this region have reached a level well below the so-called "Biomass Precautionary Approach Reference Point."

Severe overfishing by trawlers has reduced cod populations to the lowest levels ever recorded! Although closures of these fishing grounds may give the cod a chance to recover from the effects of overfishing, the measures may not be enough to save the cod fishery, since populations have now dropped so far below historic levels of the past that little is known about the dynamics of such severely depleted populations.

In 2001, the ICES had already warned the European Commission and Environmental Union governments that the spawning biomass of cod in the North Sea had reached a new historic low, and that there was a high risk of the entire fishery collapsing. Accordingly, as part of an emergency recovery plan to protect the juvenile codfish, a major section of the North Sea was closed for a 10-week period in February, March, and April of 2001. Catch quotas were also set at 50 percent of the previous year's figures. As of yet, there is still no evidence to indicate that the North Sea cod population has recovered.

Codfish are also caught as by-catch in other fisheries such as haddock, whiting, flatfish, and shrimp. ICES considers the situation so serious, it has recommended that these fisheries should also be closed, unless they can demonstrate that they can operate without cod by-catch. In addition to cod, stocks of several other commercial fisheries are below safe critical population limits in the regions under study.

ICES blames the use of large pelagic deep sea trawls, which tend to capture all marine life in their paths, for the sharp decline in the cod catches. In the past, smaller trawlers that formerly dominated the cod fishery, fished for spawning cod in the spring, and for juvenile cod in fall and winter. With the advent of the pelagic fishery in the summer, a whole new season for cod fishing has effectively been created.

On the western side of the North Atlantic, the collapse of the cod fishery of the east coast of Newfoundland in 1992 forced the Canadian government to close down the fishery completely, resulting in the loss of over 40,000 jobs. Prior to the mid-1950's, this bountiful Northwest Atlantic Cod Fishery off southern Labrador and the east coast of Newfoundland produced an annual catch of approximately 250,000 tons a year for over a hundred years. From 1956 onward, the smaller fishing vessels that had traditionally fished these waters were replaced by large factory fishing-processing ships. For the next two decades, factory trawlers from many nations of the world were allowed to legally fish within 12 miles of the shoreline of the eastern Canadian and New England coasts. In 1976, both the U.S. and Canada passed laws extending their national boundaries to 200 miles from the coastline, thus relegating foreign factory trawlers to the open seas. Canadian fishing fleets subsequently took over the region, and up to the mid-1980's over 200,000 tons of northern cod were landed every year. In 1992, the critical commercial limit for Newfoundland was reached, and Canada finally closed the area to fishing entirely.

Due to the failure of past measures in reducing fishing catches to allow for rebuilding of fisheries stocks, ICES scientists have recommended that even more stringent regulations be implemented in areas of the North Sea, Irish Sea, Scotland, and the waters which separate Norway, Denmark, and Sweden.[12]

Earth's Oceans Awash with Pollution

A recent United Nations Environmental Programme (UNEP) Report urged world governments to mount a concerted effort to reduce wastewater emissions, and thus the numbers of people at risk due to lack of basic sanitation facilities. The report stated that hundreds of millions of the world's coastal residents are presently at risk from wastewater-related problems.

According to UNEP Executive Director Klaus Toepfer, "Lack of adequate sanitation has been emerging as one of the biggest threats to human health. It is estimated that the global economic burden due to ill-health, disease, and death related to pollution of coastal waters is running at $16 billion a year." Toepfer stated that one approach to dealing with this problem is to "...set realistic but ambitious wastewater emission targets." He suggested that in order to achieve effective resolution of the problem, these targets should be linked to timetables, which stipulate when specific targets should be met.

The UNEP Report, which detailed the global threat to coastal populations and the environment from untreated sewage, indicated that the oceans of South Asia are at the highest risk of pollution, due to the 825 million people who live in that region without basic sanitation facilities. Levels of untreated sewage in South Asia's coastal waters are the highest in the world—a factor which increases the risk of shellfish contamination, and the chances of toxic algal blooms which create mass kills of fish and wildlife. Fragile marine ecosystems, such as South Asia's coral reefs, are also becoming increasingly stressed from the high levels of nutrients and suspended solids associated with the massive wastewater discharges.

East Asia is the second highest region at risk, an area where 515 million people (25 percent of the population) have no access to adequate sanitation facilities. The main objective of the UNEP report was to assist the UNEP Regional Seas Programme in taking effective action toward achieving the objectives set previously at the World Summit for Sustainable Development. In other words, to cut in half by the year 2015 the number of people that presently lacks access to adequate sanitary facilities. The problem is intensified, due to the fact that 40 percent of the world population lives in coastal regions, within 40 miles or less from ocean coastlines.

The third most polluted area is the Northwest Pacific Region, where some 414 million people are without access to adequate sanitation. The remaining regions and peoples which the UNEP report designates as at risk from sewage pollution, listed in descending order are: Western and Central Africa (107 million), the Southwest Atlantic Region (45 million), the Greater Caribbean (34 million), the Mediterranean Region (26 million), the Red Sea and Gulf of Aden (21 million), East Africa (19 million), the Persian Gulf (18 million), the Black Sea (14 million), the Southeast Pacific Region (12 million), and the South Pacific (2 million).

UNEP officials consider Wastewater Emission Targets as instruments for prioritization, resource allocation, and progress reporting in the process of achieving the global objectives, which were previously agreed upon at the World Summit. According to Cees Van de Guchte, Senior Program Officer for the UNEP Global Programme, "One additional target, which we believe is doable at the global level, is to have a minimum of 20 percent of coastal cities implementing sustainable and environmentally sound water supply and wastewater treatment systems by 2012." The longer-term goal is to provide safe drinking water and adequate sanitation facilities for all the people of the world by the year 2025. Says Van de Guchte, "Some experts estimate that this would cost $180 billion a year—two to three times more than present investments in the water sector. It may seem high, but the benefits in terms of disease reduction and dramatic environmental improvements to the coastal and marine environments are also high."[13]

**If we are to survive as a species on this planet,
let us be forewarned
that our survival depends
on how well we protect this fragile Earth,
as it is a place for each of us
to live, breathe, and bask
in the brilliant diversity of Life.**

~ Selma Brackman, Executive Director, War and Peace Foundation.

Pollution Kills Thousands of Children

A newly released study by three United Nations agencies concludes that every day, some 5,500 children around the world die from disease related to polluted air, water, and food. This report depicts the deadly threat posed by environmental degradation to the most vulnerable segment of our society—the children of Earth!

According to the UN report entitled, "Children in the New Millennium: Environmental Impact on Health," environmental contamination is responsible for a number of diseases, which include diarrhea and acute respiratory infections—two leading causes of child mortality. The study notes that, despite improvements made during the past decade in children's health and the environment, thousands of children continue to die daily from pollution-related diseases.

In the words of Carol Bellamy, Executive Director of the United Nations Children's Fund (UNICEF), "Children are healthier today. There is more access to clean water, but these disturbing figures show that we have barely started to address some of the main problems. Far too many children are dying from diseases that can be prevented through access to clean water and sanitation."

The report was presented in conjunction with a three-day UN General Assembly Special Session on Children in New York. The conference was attended by over 60 government leaders and 170 national delegations. The main objective was to move children to the top of the global agenda, by making a greater investment in children's essential social services. One of the primary goals was to increase basic access to hygienic sanitation facilities, and safe, affordable drinking water. The report identified several environmental problems that tend to affect children directly, including high levels of toxic chemicals, and the degradation and depletion of natural resources. For example, environmental lead contamination (mainly from leaded gasoline) has been shown to cause permanent neurological and developmental disorders in children.

The report also brought out the fact that millions of children that work in agricultural-related activities are at high risk from agricultural poisons. Children also tend to be disproportionately vulnerable to such environmental effects such as global climate change, depletion of the ozone layer, and loss of biological diversity. According to the World Health Organization (WHO), nearly one-third of global disease incidences can be directly attributed to environmental risk factors. Over 40 percent of this group represents children under the age of five. The single major factor that contributes to these diseases is *Malnutrition*—a condition that affects some 150 million children worldwide by weakening their immune systems. Malnutrition and Diarrhea tend to form a destructive cycle in children. Organisms that are responsible for diarrhea tend to damage the walls of the digestive tract, and prevent the proper absorption of food, thus heightening both the problems of malnutrition and vulnerability to disease.

According to Dr. Gro Brundtland, Director General of WHO, "People are most vulnerable in their youngest years. This means that children must be at the center of our response to unhealthy environments." The report highlights the fact that the public has generally remained unaware of the special vulnerabilities of children to environmental health risks. In response to this problem, UNEP Executive Director Klaus Toepfer called for international action to raise the general public awareness of the problem. He said, "I am convinced that we need to elevate children's environmental health issues on the international agenda, both through the General Assembly's Special Session on Children, and then the World Summit on Sustainable Development. We should recognize that realizing children's rights and managing environmental challenges are mutually reinforcing goals." The report stresses the need for support at the national level, with the primary focus being directed toward home, school, and community environments, where children live, work, and play. Within this context, the shift to unleaded fuel in many countries has been successful in helping eliminate lead as a source of environmental pollution.

In the United States since the 1970's, the occurrences of childhood cancers, learning disabilities, autism, diabetes, and abnormal penile development have risen dramatically. Mounting evidence suggests that these abnormalities are linked with exposure to hormone-disrupting chemicals. These persistent and hidden pollution agents are synthetic chemicals that control the way an organism develops and functions. According to Dr. Theo Colborn, Director of the World Wildlife Fund's Wildlife and Contaminants Program, "What is especially troubling is that children are exposed to these chemicals in the womb and shortly after birth—periods of rapid development." To help counteract this hidden threat to America's children, a bill has been introduced into Congress which would authorize up to $500 million for the National Institute of Health Sciences (NIEHS) to conduct an intensive five-year research program on hormone disruptors. The agency would be required to provide public information on those hormone disruptors which pose the greatest threats to humans and the environment. In the words of Dr. Colborn, "This legislation is long overdue. Not one chemical in use today has been adequately tested for its ability to undermine the construction of children's bodies and brains. There is an urgent need to support innovative research designed to identify hazards that traditional toxicology has missed."[14]

METASTRATEGY I

A UNIQUE TRANSFORMATIVE
EDUCATIONAL SYSTEM
FOR THE THIRD MILLENNIUM

The most crucial solution
to the current educational shortcomings
is to incorporate
new educational technologies
that raise life spontaneously
to be in accord with Natural Law,
and that directly unfold
the full creativity and intelligence
of the student.

~ John Hagelin—1998 *Manual for a Perfect Government,* p. 121.

Over thirty years ago, futurist, Alvin Toffler, in his best-selling book, *Future Shock,* had this to say regarding traditional educational systems:

> What passes for education today, even in our 'best' schools and colleges, is a hopeless anachronism. Parents look to education to fit their children for a life in the future. Teachers warn that lack of an education will cripple a child's chances in the world of tomorrow. Government ministries, churches, the mass media—all exhort young people to stay in school, insisting that now, as never before, one's future is almost wholly dependent upon education.
>
> Yet, for all this rhetoric about the future, our schools face backward toward a dying system, rather than forward to the emerging new society. Their

vast energies are applied to cranking out Industrial Men—people tooled for survival in a system that will be dead before they are.[15]

To help avert Future Shock we need to create an entirely new "Super-Industrial Educational System." Such a new and future-focused educational system, created especially for Third Millennium thinking and action, is desperately needed for the survival of the human race! For any such new system to be effective, we must leave the past behind, and reach out into the future—searching for creative new objectives and effective teaching methods.

Such a revolutionary new and futuristic educational system already exists. It is called *The University of the Future Project*[R] This unique educational program is both innovative, and transformational—with special emphasis being placed on the complex interrelationships between humans and the environment. The program, which emphasizes life-related activities, is designed to function effectively as a co-evolutionary catalyst, by working to develop the ideal consciousness necessary for survival and success in our emerging third-millennium society. This new educational system is part of a master paradigm called *Future-Science Technology*[R], a unique new concept, developed over the past twenty five years at the Arcos Cielos Research Center in Sedona, Arizona.

Essentially, the University of the Future Project is a new upgrade of our "Educational Brainware," a set of tools for bringing *The Art of Teaching* and *The Art of Learning* back into the heart of the educational process. Although the system was originally created as a design concept for a new type of futuristic university, it is perfectly adaptable to any educational level or system. Since the principles are flexible, they can easily be integrated into any educational level from kindergarten to university. Furthermore, the concept is also ideally suited for professional career training and lifelong learning programs.

The University of the Future program features "Sets of Opposites," which function synergistically when they are integrated into practical educational programs. Such "Oppositional Pairs" would combine ancient and modern technologies, Eastern and Western worldviews, conventional and alternative scientific viewpoints, local and global perspectives, and physical and subtle-energy methodologies.

From a future-science technology perspective, whenever the energies of two such opposites interface, a new and powerful "Creative Force" is generated. For example, in ancient times, the meeting of two cultures or religious groups often resulted in bloodshed and violence. Once these violent clashes ended, however, stable periods of cooperation, creativity, and cultural exchange would often prevail.

Within the spirit of New Millennium Enlightenment, traditional ego-centered ideological differences between our various global factions must be set aside—to be replaced with productive "Think-and-Do Tanks," and cooperative international programs, which would mutually benefit everyone concerned. This type of enlightened "Win-Win" approach to conflict resolution has tremendous potential for producing entirely new systems and solutions for most of the human-created problems which presently threaten world society and the Global Biosphere. When the differences between opposing ideologies and institutions can be tempered with wisdom and understanding, the resulting creativity and productivity can be forged into powerful new concepts and paradigms. These dynamic new paradigms would incorporate the "Best of Both Opposites." The practical solutions thus generated would have the synergistic capabilities for transforming Marshall McLuhan's "Global Village" into an enlightened, and ecologically sustainable "Global SuperNation," a planetary society which would exist within a new context of social justice and creative evolution for the mutual benefit of humans and the natural environmental systems of planet Earth.

The University of the Future Concept

Traditionally, education has tended to focus too much on past events, thought-forms, and time-frames, which were products of Industrial-Age thinking. In the words of Bennett W. Goodspeed, "Education should be focused primarily on re-capturing facilities that have been allowed to go astray in the stress of life."[16]

The linear, analytical, left-brained teaching methods of tradi-tional education over-emphasize the memorization and parroting back of isolated facts, rather than the more realistic, practical, and creative applications of these facts to real-world situations. In order to effectively bridge the gap between traditional education, and our new digital information society, a renewed infusion of enthusiasm and creative energy is required. Support from gov-ernment, corporate, and private sectors is also essential, to effec-tively transition education into a framework which is relevant to our emerging Third-Millennium culture.

The University of the Future program offers unique, educa-tionally efficient and cost-effective alternatives to the expen-sive, time-consuming, and socially restrictive inefficiencies of traditional educational methods. The new system provides the ultimate in personalized education, with major emphasis being placed on *learning how to learn.* Each student is provided with a set of basic life-management skills, designed to guarantee suc-cess and survival in real-world society. In addition to a balanced foundation of basic educational skills, learning is career-focused (i.e. personalized) in accordance with each student's unique abili-ties and aspirations.

The University of the Future System creates an educational environment which is both *supportive* and *synergistic.* Learning programs emphasize results-based experiential learning, which is strongly focused on leading-edge environmental programs. Within this context, John Gardner, in his excellent book, *Self-Re-newal,* states, "All too often we are giving our young people cut flowers, when we should be teaching them to grow their own plants. We are stuffing their heads with the products of earlier innovation rather than teaching them to innovate."[17]

The University of the Future program offers its students and teachers a variety of new adventures in self-discovery and transformative development, which occur within a Neo-Renaissance learning environment—an environment in which cooperative, synergistic relationships are formed between students and instructors. The Program also serves to enrich intercultural relations—with students sharing their cultural worldviews to generate new ideas and perspectives. "High-Tech" is artfully combined with "High-Touch," and information and communications management skills are embedded in all areas of study.

Other primary objectives of the University of the Future program include, developing the core self, attaining maturity, recognizing and acquiring wisdom, learning to practice humility and compassion, practicing support and synergy (as opposed to Darwinian competition), and developing the intuitive senses in new and adventurous ways.

The University of the Future concept regards the past as an exemplary resource to be mined for its wisdom and experience; personal resources that can be applied to virtually all aspects of personal and professional life. The majority of learning, however, is focused on present skills, events, and situations—with special emphasis on the impacts and implications of our present activities and actions on future human society and the Global Biosphere.

The Evolution of Communications Fluency

In the course of social evolution, human culture has passed a series of milestones, which include spoken and written language; the latter at first limited to priests and professional scribes. The invention of the movable-type Gutenberg Press marked the advent of printed mass-media, and a corresponding social perceptual shift from a right- to left-brained thinking modality. A progression of new and more efficient communications devices followed. These included the typewriter, teletype, telegraph, telephone, radio, television, fax machine, and computer. The integration of these "Gateway Inventions" in the social fabric was followed by the installation of transcontinental cables, and eventually wireless and satellite networks. The Global Internet not only embodies the latest digital communications technologies, but could also be said to represent the most recent electronic personification of the Earth's planetary consciousness field (i.e. *Gaia*).

In our modern world of digital communications, the following ancient oriental principle can be expanded to a new level of significance: "If a picture is worth a thousand words," a video (or movie) is worth a thousand pictures. It would thus follow, that a direct experience must be worth a thousand videos. In accord with this line of thinking, all students within the University of the Future system would be encouraged to become proficient in "Basic Communications Fluency." Skills in this category would include digital photography, videography, digital photo-manipulation and video editing; in addition to fluency in conventional written and verbal communications skills. Related subjects might include videoconferencing, e-business, and electronic presentation techniques.

"Knowledge Space"

A new anthropological space, the Knowledge Space, is being formed today, which could easily take precedence over the spaces of Earth, Territory, and Commerce that preceded it.

~ Pierre Levy, 1997—*Collective Intelligence: Mankind's Emerging World in Cyberspace.*

CyberLearning

CyberLearning integrates the technologies and resources of computer sciences, telecommunications, digital imaging, and distance learning. The global Internet and its associated technologies, have endowed CyberLearning with the capability to become an educational resource of unprecedented power and evolutionary potential!

As an interactive educational tool, distance learning is financially efficient, and adaptable to any working or parenting schedule. Distance learning is designed to be self-paced, and can thus be pursued on either a part- or full-time basis. Distance-ed programs can be set up in conjunction with local educational centers or drop-in technical labs, where computers, hardware, software, and live instructors are available 24 hours a day. Distance learning is a powerful new educational resource, and an ideal tool for implementing and enhancing the home-school experience. It is also a technology which is ideally suited for lifelong learning programs.

Additional aspects of CyberLearning should include digital information management, which combines telecommunications, internet activities, and written and audiovisual media resources— properly stored on digital storage media. Digital Information Management requires a basic familiarity with these information resources, as well as the ability to employ effective search strategies to extract, prioritize, organize, and store information in the appropriate formats. Other aspects of CyberLearning include learning how to conduct e-business, mastering basic website construction and operation, and becoming familiar with the various aspects of virtual reality. In many cases, these different communications modalities would not be taught directly, but would simply be learned as part of the process students would use to complete their educational projects.

Virtual reality in combination with other Internet communications technologies, offers tremendous potential as a learning tool. Although virtual reality environments were originally developed for military strategy and combat training, the technology offers exciting possibilities for creating "immersive educational experiences," especially when combined with "Natural Reality" (i.e. virtual field trips into natural environmental settings). Virtual reality simulators can also substitute for many aspects of learning to operate construction equipment, or piloting military aircraft, spacecraft, and deep-submersible vehicles. VR technology thus represents a cost-effective training technology, which can reduce the need for real-time operation of multimillion-dollar equipment.

An International
Digital Children's Library

A research and development team from the University of Maryland was awarded a 4.4 million dollar grant to create a free digital library for children. This free public library first opened its electronic doors on the internet in November of 2002—offering a pilot program with nearly 200 digitized books, translated into 18 languages for children between the ages of 3 and 13. By 2007 the project plans to offer a selection of 10,000 books—100 different titles from 100 different cultures. Children will be able to read, using desktop computers or miniaturized mobile devices.

The new digital library is set up to allow children to hunt for books based on the color of their covers, how the books make them feel, what types of characters they contain, and in other ways that adults might consider to be strange. One objective of the program is to test new and novel methods for navigating through the pages, using graphic, rather than textural, cues. In the words of Allison Druin, leader of the University of Maryland team, "We are developing new technologies that reflect real thinking about children's needs and cognitive abilities. If you look at existing technology for children searching on the Web, so much of it depends on reading and typing. From 4 to 8 years old, kids know exactly what they want but they have poor typing skills and a heck of a time finding it."

Another major objective of the digital library research and development team is to bring publishers, librarians, and software developers together to explore the complex issues that might arise from making books publicly accessible on the Internet.

Although publishers are generally reluctant to allow books to be publicly accessed online, several publishers such as Random House, Inc. and HarperCollins have already contributed some titles to the experimental library. Some copyrighted books presently appear in encrypted format, while others appear without copyright protection. Over two-thirds of the books are presently in the public domain, since their copyrights have already expired.

The Digital Children's Library is being designed and developed at the University of Maryland's Human-Computer Interaction Lab, with funding provided mainly by the National Science Foundation, The Institute of Museum and Library Services, and the Markle Foundation. To date, the Library of Congress has contributed 50 books. Others have been donated from national libraries in Singapore, Croatia, New Zealand, and several other countries.

In late 2002, a group of authors, librarians, and publishers from Egypt, Japan, Russia, Finland, and Germany met at the Library of Congress to discuss the five-year expansion of the Digital Children's Library. Among the unusual and unique aspects of the project is the fact that a group of children under the age of 14 are actively helping professors determine what new visual aids might help other children explore the library's contents more easily. For example, children from the Yorktown Elementary School in Bowie, Maryland have been visiting the laboratory weekly since the late 1990's, helping to develop new and unique tools for searching through the digital stacks.

It soon became apparent that children wanted to search for books based on how they made them feel—looking for "happy" books, or books that are "scary." Now, the children are rating the books on how they feel when they read them. Since the books have been scanned into digital format, covers and pages appear as digital photos on the Library's new website (http://wwwen.childrenslibrary.org).

In addition to a kid-friendly book classification system, Druin and her husband, Benjamin Bederson, Director of the Human-Computer Interaction Lab, have created a child-friendly software for viewing the digital books, that offers three different reading options: In the first option, the reader is presented with a traditional one-page view, which resembles a real book. In the second option, a comic-strip view appears, showing the book's pages as thumbnail images on a strip. The third option is a more advanced spiral view, which depicts a series of thumbnails of pages twisted into a spiral. With the last two options, kids click on any thumbnail image to zoom in and see a magnified view of that particular page.

Initially, there was some resistance from parents and educators as to whether or not kids would be willing to read from a computer screen. Now, however, according to Druin, "Nobody thinks twice." She predicts that one of the major impacts of the new digital library is that it will empower children who would otherwise be dependent on adults for their learning experience. She says, "The goal is to put kids back in control of some small piece of their life, so they get a sense of a'ha. I can do this. This is mine." It is hoped that the new International Digital Children's Library will serve to make quality children's literature more accessible across cultural lines. Says Brewster Kahle, founder of San Francisco's Internet Archive, and co-creator and web host for the Children's Digital Library, "There is no greater responsibility on a generation than to put the best of all human culture within reach of its children. That is fundamentally what this project is about."[18]

Tele-Robotic Virtual Surgery

As high-tech virtual simulators continue to evolve, their applications will become more common in areas such as medicine and environmental sciences. In medicine, human-controlled robotic Manipulators already *surpass* the capabilities of direct surgical contact between patient and surgeon. Telerobotic manipulators have recently become so precise, that human heart bypass surgery is also being performed using endoscopic techniques. Since only small openings are necessary to insert the endoscopic tubes into the chest cavity, patients are able to return quickly to their normal daily routines after a short hospital stay for routine post-surgical observation.

In a typical operating room, the surgeon seats himself at a robot control console, which is fitted with both foot and hand controls for the surgical instruments. Normally, this console is located a few feet away from the patient. The surgeon receives tactile feedback from variable resistance controls, and visual feedback from a video monitor, which can shift views via voice commands, and provide real-time feedback of the internal surgical procedures via tiny video cameras inserted in the endoscopic tubes. Although these procedures are routinely accomplished with the surgeon in the same room as the patient, a team of doctors in New York recently performed Robotic Gallbladder Telesurgery on a patient in Strasbourg, France, in what was christened, "the world's first complete long-distance operation." Watching the operation on computer screens in New York and using controls connected to sensors, high-speed signals traveled 4,000 miles across the Atlantic Ocean at the speed of light through fiber-optic lines to robotic manipulators in France, which carried out the New York doctors' commands. To honor the occasion, the doctors christened the 54-minute successful procedure, "Operation Lindbergh," in honor of aviator Charles Lindbergh's solo flight across the Atlantic.[19]

VideoGaming Technology as an Educational Tool

Video gaming technology has been responsible for placing interactive virtual reality directly into the hands of the general public. This technology offers tremendous potential as a learning tool. It can also be valuable for training the reflexes and mental-reasoning faculties. Although initially designed for children, computer and video-console games have recently emerged as a significant adult pastime. The variety of games runs the gamut from tactical military warfare, to building cities, governing kingdoms, and terraforming planets. Other "Sims" involve managing virtual families, or raising virtual "cyber-pets," with each one having its own unique personality and requirements for food, water, and affection. Such digital entities have the capability to grow, learn by trial-and-error, reproduce, and die. Although the transformation of videogaming technology into learning technology is still in its infancy, the compression of an entire interactive virtual world into a single computer disk is undeniably one of the greatest miracles of our times!

At the time of this writing for the second edition (2009) "The Sims" is the world's most popular computer game. It is also very popular with women (http://www.thesims.ea.com). "Second Life" is billed as "The world's largest virtual community," and has attracted a following of enthusiasts of all ages (http://www.secondlife.com). A complete search of "Sim Computer Games" will reveal a variety of virtual computer worlds which can prove to be delightful exercises for challenging the creative imagination of people of all ages.

Very Distant Learning

The advent of wireless and mobile technologies may eventually give a whole new meaning to the term, "Distance Learning." According to Associate Professor, Curtis Bonk, at the Indiana University School of Education, advances in communications technologies will eventually allow a professor on the Moon, or a robotic rover on Mars to conduct virtual field trips with students anywhere on Earth—or even in Outer Space. Says Curtis, "Interplanetary chats and guest experts from Mars, the Moon, and space stations will be typical educational activities by 2020."

New wireless online learning technologies will essentially help free both students and instructors from the restrictive confines of the classroom. In this type of future learning scenario, freelance instructors and guest lecturers could be available at any time and in any location, thus allowing students to design their own curricula and have personalized "Artificially Intelligent Tutors," integrated into their educational software applications.[20]

The Global Internet and "Webucation"

The Global Internet has become a major transformative force for Third Millennium society. This modern technological miracle effectively collapses both time and space, allowing anyone to communicate with other individuals located anywhere on the planet—or even in outer space. The Internet never sleeps. It functions tirelessly, 24 hours a day, seven days a week. Data and information stored in cyberspace can be accessed at any time, from any location, and shared with other individuals. As a "Planetary Equal-Opportunity Communications Medium," the Global Internet dissolves social, national, ethnic, and economic barriers. There is also evidence to suggest that the internet has significantly increased energy efficiency, and helped to reduce atmospheric pollution. As the story goes, it seems that electric commerce, e-mail, video-conferencing, and distance learning have served to eliminate a significant amount of physical travel, with corresponding increases in the efficiency of these same business enterprises.

"Webucation" also implies that the Global Internet can function as a "Conscious Entity." To cite just one example, through a process of trial-and-error, the "Website" has evolved as "The Human/Artificial Intelligence Interface of Choice"— the "Choice" being mutually agreed upon by both biological and digital entities. From a future-science technology perspective, a website could thus be described as, "A self-aware communications interface between humans and the global consciousness field." By its very nature, a website has the innate capability to adapt and evolve, according to the digital inputs and interchanges which occur within its boundaries. The "Website" thus seems to have developed into a "Cyberspace Entity" in its own right. Websites continue to become increasingly sophisticated; and this sophistication continues to increase at a breakneck pace. Websites are constantly improving their efficiency. They communicate via audio, animation, and real-time video, and continue to incorporate the latest hardware and software technologies into their structure. Considering the transformations which have already occurred, it seems only a matter of time before the basic website concept shifts from a two-dimensional entity into a three-dimensional cyber entity, which embodies both virtual reality and holographic technologies, eventually incorporating immersive virtual adventures, which will rival the "Holodeck" on the Starship *Enterprise.*

Home-Schooling: a Powerful Educational Tool

- **Educator Dr. Howard Richmann found that home-schooled students in Pennsylvania produced averages of 86th percentiles in Reading and 73rd percentiles in Math (The national average is the 50th percentile).**

- **Dr. Larry Shyers, observing children in free play and group interactions found that conventionally schooled children had significantly more problem behaviors than did their home-schooled counterparts.**

- **Dr. Gary Knowles of the University of Michigan studied a group of home-educated adults. None were unemployed, and none were on welfare! Ninety four percent of this group felt home schooling had prepared them to be independent and 79 percent felt that home-schooling had helped them interact with individuals from different levels of society.**

- **Between 1.5 and 1.9 million students in grades K through 12 were home-schooled in 2001-2. Home educators and their families are not dependent on public tax funding. In the State of Oregon alone in 1997, home schooling was estimated to have saved taxpayers at least 61 million dollars annually.**

  ~ *The McAlvany Intelligence Advisor,* November 2002, p. 16.

Creating a Synergistic Learning Community

In order to create a synergistic learning community a futuristic learning environment must be developed that can function as a co-evolutionary ecosystem—a learning community that nurtures both the heart and spirit. Key aspects of this learning community would include spiritual growth and development, and the creation of an evolvable group consciousness. Other key elements include: a regard for the sacredness of the natural environment, and respect for the personal space, feelings, and individuality of others.

In order to bring the more esoteric aspects of the University of the Future Project into practical reality, a basic set of body/mind development skills have been included in the program. Such skills include food preparation, proper eating habits, advanced nutritional education, physical development and recreation, yoga and the martial arts, and Paraphysical Fitness®, a program designed to integrate an individual's physical and mental functions with the higher consciousness centers, thus creating a balanced, self-sufficient, and integrated person.

Another often neglected aspect of holistic education is a practical set of life management skills, critical for survival, success, and happiness in the day-to-day activities of real-world living. Key life management skills would include: proper manners and social behavior, appropriate dress and appearance, time and financial management, career development, mate selection, parenting, family management, and reproductive responsibility. The main objective of this program would be to establish a self-generated code of ethical behavior by stressing the importance of honesty, integrity, commitment, and a thorough understanding of a karma-based value system—with "Karma" being defined as, "The ultimate example of self-responsible behavior" (i.e. What goes around comes around). Christianity's Golden Rule or the Buddhist Eight-Fold Path would serve equally well as an appropriate behavioral model for third millennium society.

Village Internet-on-Wheels

Recently, village residents in India, who do not use telephones, or have never even seen a computer, can now have their first "Internet Experience," thanks to an innovative new project, funded by a seed grant from a company called Digital Partners. The villager's new "Computer on Wheels" is essentially a motorcycle which carries a laptop computer, and is driven by a computer technician. This innovative combination of "Practical Transportation" and "High Technology" gives rural villagers the opportunity to view web pages on the computer screen that have been downloaded from the Internet.

This pilot project for creating a mobile Internet service was recently begun in the Talangana region of the southern Indian State of Andhra Pradesh. Initially, the pilot project is scheduled to run for a year. If successful, the program will be expanded to cover the entire state. As explained by Satish Jha of Digital Partners, "Much like the post office, where the postman delivers letters once or twice a day, we are delivering the internet to people once or twice a day." The Seattle-based non-profit organization sees its Internet-on-Wheels Project as a way to involve India's millions of rural citizens in the Internet Revolution.

Says Project Director Jha, "Why should a whole section of the population who don't have telephones, who don't have electricity be left behind? Seventy percent of the villagers do not have access to telephones or electricity, so how can they use computers? We need to find ways of taking the computer to them."

Since the villages have no Internet connections, relevant Web pages are first downloaded into a laptop computer. A computer technician then drives the motorcycle to the villages, usually twice a day. Villagers can request certain specific services, like crop prices for regional markets, or the latest local news for their area. Says Mr. Jha, "There is an element of curiosity. As soon as they hear the motorbike and know the Laptop-on-Wheels is coming, 50 to 100 people will collect around it." Jha likes to compare the arrival of the laptop to the early days of movies, when local villagers would crowd around the screen to steal a peek at the moving picture shows.

Although the Computer-on-Wheels project is still in its early stages, the organizers are hoping it will prove to be a viable way to overcome the lack of a proper communications infrastructure in the rural areas of India.[21]

Computer Village in the Clouds

Nepalese education pioneer, Mahabir Pun, is determined to break the cycle of poverty in his native mountain village of Nangi, by bringing it into the computer age. In 1996, one of Pun's professors helped him set up a simple website, which provided basic information about his school and village. This website effectively connected the students to the outside world and also attracted the attention of people in other countries, who then began to help by sending books, teaching materials, computer parts, and money.

Used computers and component parts soon began arriving from Australia, the United States, Malaysia, and Singapore. Pun and his students set up the computers, and assembled others, often using wooden boxes to hold the components. Since the village had no electricity at the time, Pun installed two hydroelectric generators in a nearby stream. The school now has 15 computers for the 300 students, who come from six neighboring villages. Previously, only wealthy private schools in the city offered computer classes, while most of the students from the rural areas of Nepal had never even seen a computer—let alone operated one.

In 1966, Pun was finally able to get a phone line into the village; however transmission quality was not good enough for an Internet connection. So, about once every month, a group of students and teachers walk for a full day to the nearest city, where Internet service is available for them to communicate with other individuals around the world. Although the group would like to establish a satellite Internet connection to their village, for the time being, this option is too costly.

With Internet access in the village, the website could be used for raising money, which would help provide quality education for all the local children. It would also be used to provide information about Nepalese culture to children all over the world, and invite volunteers to visit the village. Educational pioneer Mahabir Pun eventually hopes to establish a university in his village, where computer classes can be held for the local students. This would open the way for them to become computer programmers themselves. They could then produce software for big companies in other parts of the world, without leaving their home villages. The key for success with this plan, of course, is Internet access.[22]

"Consciousness-Based Education"

**Education should unfold full creativity
and higher states of consciousness,
to produce citizens capable of fulfilling
their highest aspirations,
while contributing maximally
to the progress of society.
By harnessing the nation's greatest resource—
the unlimited creativity of its citizens—
effective education can ensure
National Prosperity, International Competitiveness,
and a Leadership Role in the Family of Nations.**

~ John S. Hagelin, *1998—*
Manual for a Perfect Government, p. 120.

Futuristic Education and
Subtle-Energy Management

One fundamental aspect of the University of the Future program involves the understanding and practical applications of subtle-energies in all aspects of our lives. By achieving an understanding and functional knowledge of subtle-energies, we can learn how to operate at the interface between the physical and quantum realities, to enhance and magnify the basic capabilities of the human mind. Considering the recent integration of selected alternative and conventional medical practices, this type of integrated scientific approach is already becoming a practical reality. Similar integrated approaches in other areas of science, education, and technology are already beginning to yield correspondingly dramatic innovations in their respective areas.

Subtle-energy phenomena which are already integrated into the University of the Future Programs include: Earth energies—an understanding of natural Earth energy currents and their effects on the human body, and Supersensonics, the awareness, development, and integration of our natural intuitive abilities into business, science, and daily lives. Other subtle-energy aspects of the program include Environmental Energy Enhancement—the magnification and focus of naturally positive energies, achieved by combining ancient traditional sciences such as Feng Shui and Geomancy, with modern technology. Subtle energy awareness, a fundamental aspect of both consciousness development and Paraphysical fitness, is achieved through special exercises, designed to develop concentration, enhance mental efficiency, and stimulate creative thinking. Martial arts and related esoteric systems such as Yoga, Tai Chi, and Chi Gong all function correspondingly to augment and integrate human mental and physical abilities, and to expand and enhance the intuitive senses.

**Anything
that continually creates itself
is a Living Entity.**

~ Kenneth J. Cox, 2003—"Outer and Inner Space Explorations,"
Presentation prepared for NASA
Scientists and Administrators.

Planetary "Self-Consciousness"

When television pictures of our planet were first beamed back to Earth from space, this event was said to be the beginning of self-awareness for planet Earth—the point at which the "Planetary MindField" first achieved self awareness. In his book, *Unknown Man, the Mysterious Birth of a New Species* author Yatri described this transformation as follows:

> **With the addition of new artificial intelligences to the network, and the developing richness and quality of the flowing information, Marshall McLuhan's 'Global Village' has changed into the 'Global Brain'—an entirely new kind of Consciousness and a new kind of Planetary Species.** [23]

The invention and rapid deployment of Communications Devices such as the telegraph, telephone, radio, television, VCR, fax, interactive Internet, and earth-orbital satellites may have indeed created a new "Self-Aware Global Intelligence." Thus, metaphorically speaking, the one-room schoolhouse—the educational foundation of rural America—has apparently metamorphosed into something which was inconceivable only a century ago. The computer itself has now become an educational portal to the Global Brain—essentially, a "Global Schoolhouse-in-a-Box."

The University of the Future Concept embraces the development of high-energy, intellectually stimulating, and emotionally supportive learning environments, designed to teach and nurture each individual student. Within these immersive experiential environments, students would undergo adventure learning experiences, designed to develop self-reliance, creativity, spiritual sensitivity, the evolution of consciousness, and the pursuit of happiness. For a paradigm to be both dynamic and timeless, it must also be flexible, evolvable, and grounded into practical reality. These objectives are fundamental to the University of the Future Project—an educational system whose embedded elements are appropriate for meeting the educational challenges of both the 21st century, and the Third Millennium.

Basic Strategies for a Futuristic 21st Century Educational System:

1) A new educational system for the Third Millennium should be based on environmentally focused direct-experience learning, taught within an educational environment that replaces the Darwinian concepts of conflict and competition, with the concepts of creativity, cooperation, and consciousness development.

2) Effective communications competency and digital fluency should be emphasized at all educational levels. Subjects would include: computer literacy, distance learning, digital photography and videography, media presentations, website design and construction, and the fundamentals of electronic business.

3) High-tech should be artfully combined with high-touch. Caring and sharing between students and teachers would be the norm—resulting in the emergence of powerful new learning systems, which would continue to evolve and be socially relevant for the future.

4) Group-consciousness development and a sense of community (i.e. belonging) should be encouraged throughout the educational system. Students would be brought together to experience creative thinking and synergistics for the benefit of their own learning community—while remaining ever-mindful of the greater interrelationships of humans to all life on Earth.

5) A Neo-Renaissance Approach to Learning should be emphasized. This approach would focus on mentoring and hands-on, project-centered learning. Each student would be provided with a working knowledge of basic skills, which could be used throughout their lives. Self-reliance, creative thinking and problem solving would also have high priority in the educational programs.

6) Most subjects should be "Career-Related," in that they would be designed for optimal success, survival, and happiness

in the real world of personal and professional activities and interactions.

7) Consciousness technology (i.e. the development of each individual's innate intuitive abilities) should be incorporated into all aspects of the educational programs. A heightened awareness of the various aspects of consciousness technology will be key aspects for survival and success in our rapidly shifting and evolving Tri-Millennial society.

METASTRATEGY II

A SUSTAINABLE GLOBAL
RESOURCE MANAGEMENT PLAN

**The species seems to be
eating its way across the planet,
gobbling up in a few short decades
resources which took millions of years to accumulate.
We have introduced entirely alien ecosystems
into the organic totality
of the planetary system
and are threatening to destroy
the whole biological framework by doing so.**

~ Yatri—*Unknown Man: The Mysterious Birth of a New Species.*

Environmental Destruction of Earth's Land Masses

Director of the Millennium Institute, Gerald O. Barney, had this to say regarding the clash between human activities and the Global Biosphere:

> **"...If we people of Earth are to avoid a massive disaster within the lifetime of our children, our most critical and urgent task is to bring forth a transformed vision of progress, one of sustainable and replicable development."[24]**

The oceans, land masses and atmosphere of planet Earth are the "Environmental Commons" for the global community, since whatever happens in one area on earth, theoretically affects all other parts of the Global Biosphere.

Throughout the past millennia, the forests and vegetative areas of our global land masses have been drastically altered by human activity. This wanton destruction continues every day at an alarming pace! For example, according to *World Watch* Editor, Ed Ayers, over 100,000 slash-and-burn fires are set each year in Mexico, the Amazon, and Malaysia. The world's tropical forests continue to decline by an area equal to one football field *every second.* During this process, three species are eliminated *every hour.*

Humans have also historically consumed topsoil at a frightening rate—much faster than nature can possibly restore it. Each year, some 30 billion tons of the world's topsoil are lost to erosion. According to Ayres, losing topsoil can be likened to a person losing blood—as only so much can be lost.[25]

From a slightly different perspective, it has been estimated that over 75 percent of the topsoil that existed worldwide when Europeans first colonized the New World, has disappeared. This massive-scale global devastation has caused substantial damage to the water cycle by interfering with our global forest watersheds.[26]

Biologists surveyed
by the Museum of Natural History in New York
say we have entered
the fastest Mass Extinction in Earth's history—
even faster than when the dinosaurs died.

~ Ed Ayres, 1999, *God's Last Offer,*
Negotiating for a Sustainable Future, p. 27

Symptoms of an Unhealthy Planet

Emerging diseases in marine species are indicative of the ongoing degradation of Earth's coastal ecosystems, as well as the effects of global climate change. Altered coastal biology, global warming, and climate variability all contribute to set the stage for emerging diseases, which impact a wide range of different species. Marine coastal environments are subject to increasing pressures from residential, recreational, and commercial development. Spills, leaks, and accidents associated with oil extraction and transport create additional disturbances to our coastal marine ecosystems.

Sea birds, turtles, and marine otters—animals at the top of coastal food chains—function as "Bio-indicators," since they tend to be the most sensitive to the effects of pollution and those environmental changes resulting from human activities. Coral reefs, the most diverse ecosystems on the planet, are also especially vulnerable to disease and temperature changes. Diseases and coral bleaching have already resulted in the major deterioration of coral reefs in many areas of the world.

According to recent statements by environmental experts advising the U. S. Senate, "We are witnessing the degradation of coastal marine habitats through excessive loading from farming, coastal development, animal and human waste, and the burning of fossil fuels. Extreme weather events, such as heavy rainstorms and flooding, have increased in intensity, and are projected to increase in frequency with continued climate change."[27]

> **The Oceans are a great metaphor—**
> **Everything we do on land**
> **ends up in our coastal waters.**
> **The Oceans mirror the health**
> **of our Planet.**
>
> ~ Ted Danson, President, American Oceans Campaign

20 Years to Repair Mexico's Environmental Damage?

According to Mexico's Environmental Minister, Victor Lichtinger, it will take 20 years or more of strong policies to rescue the national environment from the decades of abuse and neglect that have ravaged the country's forests and polluted its air and water. Says Lichtinger, "It will take a good number of years to resolve the most important problems. It will be about 20 years, if we pick up the pace, before we can feel that Mexico begins to be a more sustainable society."

Mexico, which ranks second only to Brazil in its rapid rate of deforestation, also faces a major national water supply crisis. New satellite studies indicate that between 1993 and 2000, Mexico lost more than 2.8 million acres of its forest cover—twice as much as was originally thought! The major ongoing threat to Mexico's forests is from poor farmers, who slash and burn the forests for farming and livestock grazing. The traditional system of subsidies simply encourages these peasant farmers to continue their wanton expansion into the forest ecosystems—planting crops and clearing trees to create new grazing lands.

Since December, 2002, Lichtinger has been instrumental in implementing a series of environmental reforms he claims will gradually function to improve the national environment. These new programs include establishing a National Forestry Commission, which has changed subsidies for peasant farmers, encouraging them to protect forested lands, rather than to cut the trees down, and a new federal grant program, designed to stimulate state and federal governments to charge consumers for the water they use. The Mexican congress has also recently passed legislation which will force logging companies to monitor and report toxic emissions. Under the new conservation initiative, federal security forces are now regularly deployed to counter illegal logging operations throughout the country.

New programs recently put into place, subsidize farmers who plant new trees, and assist in protecting the forests. Says Environment Minister Lichtinger, "We have to give them money so they can buy com, so they can live and at the same time look after the forests." In conjunction with the new subsidy programs, the newly formed National Forestry Commission has plans to reforest some 519,000 acres in the year 2002. Other priorities include developing new policies which will prevent further environmental degradation of Mexico's main river basins and their associated watersheds. In the words of Lichtinger, "The idea is, little by little, to change the trend, to lose less and less (forest) and in some critical areas, especially in protected areas, to begin recuperating."[28]

Our Global Water Crisis

One of the major consequences of human activities on the environment is that global water resources have come under intense pressures—mainly from human and industrial pollution, and especially from deplorably wasteful practices in domestic, industrial, and agricultural sectors.

According to the U.S. sponsored Third-World Water Forum, an impending global water crisis could affect one out of every three people by the year 2025! Since some 450 million people in 29 countries already suffer from water shortages, it is imperative that governments begin coordinating their efforts to develop new, resourceful methods to conserve shrinking water supplies. The Middle East, India, Pakistan, and China are all presently facing major water crises. Large-scale regional programs need to be implemented to make agricultural water use more efficient, and to reduce all forms of pollution—the major causes of water contamination.[29]

Satellite Monitoring of Global Forest Cover

Far more accurate estimates of global forest cover can now be made by combining data from NASA's new Terra Satellite, with a new method for mapping tree cover recently developed at the University of Maryland.

According to Geography Professor John Townsend at the University of Maryland Institute for computer studies, "It is essential that we know accurately how much forest cover there is on our planet, to help us conserve what is left, and how to make the best use of forest resources."

Recently, a major effort has been made by the United Nations Food and Agriculture Organization to obtain reliable statistics for global forest cover. However, they discovered that all too frequently, the figures were either not uniformly accessible or, in many cases, highly inaccurate. FAO researchers thus concluded that global forestry has many inherent problems—

simply due to confusion with the definition of "what actually constitutes a forest."

Remote satellite sensing now provides a globally repeatable and verifiable technology, which completely eliminates previous problems of local data distortions from regional institutions and agencies within each country.[30]

Polluters Can Run, but They Can't Hide

For the first time, scientists and policy makers have a means to identify major global atmospheric pollution sources. They can now track pollution as it travels through the atmosphere, crossing continents and oceans (http://www.acd.ucar.edu/mopitt/).

A new orbiting monitor called MOPITT was installed aboard NASA's Terra Spacecraft, which was launched in December 1999. MOPITT, which circles the Earth from pole to pole 16 times a day, has the capability to map pollution sources that cover areas as small as a few hundred square miles. Its sensitivity is accurate enough to distinguish differences between pollution from large metropolitan areas, and major forest fires.

Scientists at the National Center for Atmospheric Research (NCAR) in Boulder, Colorado, are integrating this new data into their existing computer model of Earth's atmospheric chemistry. This new blending approach will now enable scientists to work backwards from observation points to pinpoint the sources of the pollution.

MOPITT thus exemplifies a new and unprecedented scientific capability for scientific observations of carbon dioxide concentrations on a global scale. In addition to being a major greenhouse gas, Carbon dioxide is also an indicator of other types of atmospheric pollution. The new satellite technology thus represents a new tool to assess the effects of major urban pollution on the Global Biosphere.[31]

Marine Foreign Species Introduction

Marine foreign species introduction is not a recent phenomenon, since over the past few centuries, invasive alien species have

caused extensive damage to natural ecosystems and human economies around the world. Filling ship's holds with sea water ballast
is believed to have contributed to the translocation of many marine
species throughout the world's oceans and waterways. That some
of these organisms are now a credible threat to their new environments has already been borne out by the destruction of several
entire fisheries industries in the Black Sea.[32]

Biological invasions of the zebra mussel alone are said to
have cost the U. S. an estimated five billion dollars, since its introduction into the Great Lakes via freighter ballast in 1988. It is
now estimated that some 20 percent of all freshwater fish species
are at risk for extinction in the near future, due to foreign species
introduction. Invasive Plant Species have also already established
themselves in over an estimated 100 million acres in the United
States. They continue to spread across three million additional
acres every year.[33]

Underground Fires: A Hidden Cause of Atmospheric Pollution and Global Warming

Although hidden from the public eye, underground coal fires
are burning continuously in various locations around the world.
The United States has several dozen—most of them started by the
burning of trash in abandoned coal mines. Similar mine fires are
located in India, China, Australia, and Indonesia, where uncon-
trolled underground fires are among the wildest and most severe
on Earth.

In 1997 and 1998 Indonesia experienced a rash of uncontrol-
lable fires for several months, when millions of hectares of forests
burned—enshrouding the land in clouds of choking smoke and
haze. In 2002, the fires returned again with a renewed vengeance.
Although most fires were associated with logging operations, many
of the worst fires occurred in wild uninhabited regions, starting
on perfectly clear and cloudless days, which ruled out lightning
strikes. Researchers finally came to realize that coal seams,
burning just below the Earth's surface, are the likely culprit

in causing these otherwise unexplainable fires. As a result, the Indonesian government has initiated a program to protect its forest resources, using a newly developed technique to extinguish these underground fires.

Underground coal fires can burn anywhere from one to ten meters below the surface of the forest floor, and have been known to extend for tens of kilometers. One of the most famous underground fires began in 1961 in Centralia, Pennsylvania. It was accidentally started when someone decided to burn some trash in an abandoned coal pit. The fire somehow ignited a coal seam that extended for several miles. During the next few years the underground fire continued raging out of control, despite concerted efforts by town authorities to extinguish it with water, block the vents with concrete, and excavate the burning coal. As costs escalated into the millions of dollars, it was finally decided to relocate the entire town of 1,100 people. Experts expect the Centralia fire to continue burning for centuries.

The significance of this event is that similar attempts to put out underground fires have been repeated without success, year after year, in various locations throughout the world. For example, in 1999 a team of geologists on the Ute Indian Reservation in Colorado attempted to extinguish a coal fire that was undermining a local forest. The group spent nearly a million dollars trying to block the vents that were feeding the fire—plugging them with a styrofoam-concrete mixture, in an effort to cut off the fire's oxygen supply. For a while, this tactic appeared to be working, but several weeks later, a new vent opened up several hundred meters away from the last one, and the fire regained its original strength. Unfortunately, this type of experience is typical with most underground coal fires, since, besides sucking in air from vents that open to the surface, the fires can also bring in air through porous gravel soils. For this reason, simply pouring water into a vent, or attempting to block it with concrete, will usually prove unsuccessful.

In the early 1990's a team of firefighters from the Office of Surface Mining in Washington, DC, after many unsuccessful attempts to extinguish underground coal fires, came to the conclusion that any attempt to suffocate an underground fire was ultimately doomed to failure. The team, led by Alfred Whitehouse, developed an entirely new and better approach. First, they would create a firebreak to block the forward advance of the flames. They would then excavate the coal just beyond the leading edge of the fire. This approach effectively isolated the fire from the rest of the coal seam, and prevented it from spreading. This simple strategy proved to be successful against many coal fires in the United States—especially in the mines.

In 1997, the Indonesian government asked for a U.S. grant and a team of experts to deal with their underground coal fires. A new global initiative established at the Kyoto Protocol, for the first time, emphasized the importance of underground coal fires as a major source of air pollution. Since these underground fires constitute a major source of carbon dioxide emissions, eliminating them could be easier and more effective than the complex process of reducing industrial and vehicle emissions.

Underground fire expert whitehouse approached Indonesia's problem by first determining how many of the country's underground coal fires had already been documented. Previous surveys by a Dutch forestry research group had already located some 73 underground fires—as many as exist in the entire United States. Combining his own research with existing data, Whitehouse estimated the number of Indonesian coal fires to be around 100,000. This astonishing figure implies that underground fires represent a major global source of carbon dioxide—which well may exceed the combined CO_2 output for all vehicles and industrial plants in Indonesia! Whitehouse now feels that it is these underground coal fires that are responsible for the seemingly mysterious ability for forest fires to occur year after year—even in regions which are remote from human habitations, with no lightning storms in the area. During the long winters, when the forests remain wet, the fires apparently continue to smolder underground.

In 1998 Whitehouse and a team of ecologists, devised a simple and practical plan to teach groups of local people how to extinguish these underground fires. Together, the group was able to extinguish coal seam fires that had started that year. All that was required was a crew of men with shovels, a long thermometer for taking ground temperatures, and a water pump to cool down the coal that is too hot to excavate. Manual labor is used to dig out the coal in an arc-shaped trench—just ahead of the fire as it moves along the coal seam. Once the fire is deprived of its underground fuel source, it will simply burn itself out. According to Asep Mulyana, an Indonesian geologist, working with Whitehouse, "We have trained people from local and county governments, from mining and timber companies, and NGO's. Before Alfred's work started, people here didn't believe it was possible to put these fires out." Mulyana hopes that a small portion of the taxes from logging and mining industries can be set aside for the purpose of fighting underground coal fires, especially on the Islands of Sumatra and Borneo, where forest fires present a serious threat to the lush tropical environment.

The next logical step is to expand and coordinate this underground firefighting program up to an international scale. Scientist Johann Goldammer from the University of Freiburg in Germany plans to utilize a new satellite called "Bird," launched in 2001 by the German Aerospace Center (http://www.lsespace.com/missions/bird.php). He hopes to use the satellite's advanced thermal imaging cameras to pinpoint the Indonesian underground fires at a much finer resolution than has previously been possible. Initial images from the satellite cameras indicate that the "hot spots" from the high resolution scans correlate directly with the locations of known coal fires. In the future, Goldammer intends to use the satellite data to study correlations between the triggering of forest fires in remote areas, and the locations of thermal hot spots.

Indonesia is not the only country with a serious underground coal fire problem. In 1999 Dutch researchers conducted a survey of underground coal fires in China, using NASA's Satellite, "Landsat." Their results confirmed previous Chinese research, which indicated that coal seam fires in China could be consuming up to an estimated 100 million tons of coal per year![34]

Research strongly suggests that underground coal fires, in addition to destroying priceless forest resources, also constitute a significant source of global atmospheric pollution, which in turn exerts a major influence on global weather patterns and climate changes. With the new satellite imaging technology, plus the simple and effective strategies recently developed for putting out these fires, it would seem appropriate to create a "Global Ecological Task Force" to locate, monitor, and extinguish these fires. This type of coordinated international program could effectively eliminate this major cause of atmospheric pollution and environmental destruction within the space of a few years.

Fiber, Not Forests

America uses about as much wood, by weight, as all metals, plastics, and cement combined. The majority of American lumber is shipped to Japan, with about 40% of these trees being used for the production of paper—most of which is *not* recycled. If we continue on this present course, the majority of Earth's temperate forests could well disappear within the next few decades![35]

Considering the fact that since 1937, about half the world's forests have been cut down to make paper,[36] one practical approach to reducing our dependence on trees for fiber, is to realize that a large percentage of the trees presently being pulped for paper and building materials could be saved if we replaced them with alternative sources of annual fiber. Practical "Treeless Fiber Substitutes" can be derived from annual fibers like industrial hemp—a valuable low-cost biological resource that can be grown in almost any climate. Hemp is both fast-growing and hardy. Its high resistance to diseases eliminates the need for expensive herbicides and pesticides, as is true with commercial fiber crops like cotton, which require large amounts of these agricultural chemicals. Before the cultivation of industrial hemp was outlawed by the U.S Government, the 1913 *Yearbook of the U. S. Department of Agriculture* referred to hemp as "the oldest cultivated fiber plant," stating how the crop "improves the land," and that it yields "one of the strongest and most durable fibers of commerce."[37]

Paper from Clay?

A forgotten, ecologically friendly tree-free technology for making paper from bentonite clay was developed at the Massachusetts Institute of Technology by Dr. Ernst Hauser in the early 1940's. Hauser's unique product, originally called "Alisfilm," was named for the main ingredient in bentonite clay, a substance commonly used as a mud lubricant in commercial well drilling.

The finely ground bentonite clay is first suspended in water, allowing grit and foreign materials to settle out. The suspension is then evaporated into a jellylike mass, and chemically treated. The microscopic mineral fibers mat together during the drying process to form a film which is similar to ordinary fiber-based papers.

Paper produced by this process is waterproof and ultra fine in texture, making it ideal for printing and packaging applications. This dormant technology—elegant in its simplicity—represents an ecologically appropriate way to reduce the impact of the paper industry on our global forests. The process is covered by three U. S. patents.[38]

Paper from Pond Scum

Scientists at the University of Texas at Austin recently discovered nine species of cyanobacteria (some of the oldest organisms on Earth) that secrete cellulose, an essential ingredient for many types of materials including numerous fabric and paper products. Although cellulose is normally extracted from wood pulp or cotton, both these sources have impurities that are expensive to remove.

Since cyanobacteria are low on the food chain, they are biologically efficient, and can produce cellulose from sunlight, carbon dioxide, and water. In their natural state, however, they do not yet posses the biological capabilities to produce cellulose on a commercial scale. The Texas scientific team is thus attempting to use genetic engineering to increase the cellulose output of the cyanobacteria to the point where the process becomes commercially feasible. The idea is, that in the future, this innovative biotechnology might be implemented, so that sheets of cellulose could be skimmed off the surfaces of algae culture ponds—creating a tree-free source of paper, which would eliminate the need to harvest cellulose from the wood in our forests.[39]

World's Fastest-Growing Trees

With their thick green stalks and leaves that look like elephant ears, Paulownia Trees conjure up images which are reminiscent of the fairy tale, *Jack and the Beanstalk.*

Paulownia trees originated in Asia, but were introduced into North America around 1850, spreading in the wild from New Jersey to the Carolinas. Presently there are about 10,000 acres of Paulownias grown commercially—mainly in the southeastern and mid-Atlantic states. Former U.S. President, Jimmy Carter has 15 acres of these trees planted near his home in Plains, Georgia. These amazing trees can grow as much as 20 feet a year, reaching heights of 35-40 feet. In referring humorously to Palownia's rapid growth, Mr. Carter was recently quoted as saying, "Don't put your face over it. You may get a faceful of leaves."

The wood is light and strong, but resists shrinking or splitting, making it ideal for furniture, musical instruments, plywood, and molding. The main problem with this new renewable timber resource is that the lumber industry is basically conservative, and tends to resist new products. Also, from a commercial standpoint, there are not yet enough Paulownia trees to produce sufficient board feet of lumber for the timber industry to get seriously involved.[40]

The Test-Tube Forest

We are witnessing the dawning of a new age of bio-engineered forestry—a technology which has the potential to completely revolutionize the $750-billion-a-year global forestry industry, and to transform forest landscapes all over the world. Cellfor, Inc, a Company in Western Canada, has developed a revolutionary New technology for creating genetically engineered, fast-growing "Super-Trees."

Cellfor's super-trees begin their existence as tiny greenish-brown dots, pressed into nutrient cakes in laboratory Petri dishes. These tiny tree embryos are clones from some of the finest specimens of Douglas fir ever found in the wild. These trees, which blanket much of the Pacific Northwest, tower to heights of up to 200 feet, and are highly prized by lumber companies for their straight, strong, knot-free wood.

According to Cellfor's President, Christopher Worthy, "What happens when you plant these trees is strikingly different from what happens when you plant ordinary unimproved trees." The company's proprietary technology opens fascinating new possibilities for bypassing traditional seed orchards, by creating millions of copies (clones) from a single prototypical seed and storing them cryogenically. Traditional tree farms have never used selective breeding to improve forest trees (a technique which has radically transformed most forms of domesticated plants and animals over thousands of years). Since forest trees are not easily propagated by cuttings, forest biotechnology was considered to be impractical, and economically unfeasible.

Cellfor has developed a unique set of clonal technologies, which has the potential to increase wood production by as much as 60 percent from one generation to the next! Traditional lumber plantation production is approximately 30-40 cubic feet of wood per acre. Through improved breeding and management techniques, foresters have already been able to increase conventional forest lumber production by a factor of ten or more—even without genetic engineering. To date, the U.S. Department of Agriculture has already received applications for testing 138 new types of genetically modified trees.

Cellfor's new biotechnology techniques offer exciting possibilities for designing trees from the ground up. Forest researchers believe that within the next few years, companies like Cellfor will be able to create entirely new kinds of trees. These new "Super-Trees" would be fast-growing, knot free, and almost branchless, making them ideal for lumber and related products. Such modified super-trees, could be grown in intensively managed "Super-Forests," and could even be genetically engineered to produce alcohol, or almost any other chemical directly from solar energy and carbon dioxide.[41]

Plastic Lumber

The first suspension bridge built entirely of plastic lumber was constructed on the Hudson River Interpretive Trail in New Baltimore, New York. The bridge was designed to handle pedestrian and light truck traffic. This innovative new technology was developed by U. S. Plastics Corporation, the largest producer of plastic lumber in the world.

U. S. Plastics has two main business lines: 1) A Division for manufacturing plastic lumber, packaging, and other products from Recycled Plastic, and 2) A Division dedicated to environmental recycling services. This company's main product line includes: competitive building materials, furnishings, and industrial supplies, fabricated from recycled plastics into purified, consistent products—all designed to be artistically pleasing, and environmentally responsible.

Plastic lumber is far superior to wood in terms of durability and low maintenance. It also has numerous applications for outside decks, patios, public benches, waste receptacles, and especially for marine structures such as docks, where marine boring organisms and moisture resistance are primary considerations. In tropical areas of the world plastic lumber offers advantages as a unique termite-resistant, moisture-proof, tree-free lumber substitute. Transforming recycled plastics into versatile and durable building materials is an example of the type of "Win-Win" technology which is needed to minimize the impact of human activities on the natural environment.[42]

Tropical Rainforests: Our Precious Resources

Although tropical rainforests cover less than two percent of the Earth's surface, they are home to 50 to 70 percent of all known species—some 30 million species of plants and animals. This includes two-thirds of the world's plant species, and includes many exotic orchids, and plants with priceless medicinal value. Rainforests are thus among the richest, most productive, and most complex ecosystems on the Planet. Scientists have estimated that their destruction results in an average of 137 species being driven into extinction *every day*—or over 50,000 *a year!*

Rainforests function as important regulators for the global weather systems. Their destruction alters the global hydrological cycles, which results in drought, desertification, flooding, and soil erosion. The presence or absence of rainforests also alters the albedo (reflectivity) of the planetary surface which, in turn, impacts wind and ocean current patterns—shifting both rainfall amounts and distribution patterns.[43]

"Clearcutting" the Ocean Floors

Destructive environmental practices also occur in the world's oceans, where wasteful fishing practices continue unabated. Weighted trawling nets, dragged across the ocean floor, catch huge numbers of fish, shrimp, and shellfish. Large commercial trawlers are damaging the sea-floor environment on a massive global scale! Since trawling is hidden from public view, its destructive effects are easily overlooked. Commercial trawling nets tend to crush or bury seabed organisms, effectively destroying their food and nursery grounds—ecosystems which are critical for replenishing marine seafood stocks. Trawling has thus been compared to the clearcutting of forests. Scientists have estimated that the global trawler fleets disturb an area of ocean floor one hundred fifty times larger than the areas of Earth's forests cut each year!

In the words of marine scientist, Carl Safina, "Bottom trawling is akin to harvesting corn with bulldozers; a single pass kills from

5 to 20 percent of the seafloor animals." Twenty four hours-a-day, seven days-a-week, commercial trawlers continue their relentless destruction of Earth's seabed environments—year-round, and on a massive global scale. Suggested actions for managing this fisheries resource have included: 1) Creation of more "No-Fishing" replenishment zones, 2) Implementation of minimal-damage fishing practices, and 3) New financial incentives to aid in the development and use of "Environmentally Kinder" fishing gear and practices.[44, 45]

Wasteful Fishing Practices

Every year in the Bering Sea and Gulf of Alaska, *over 300 million pounds of fish and marine organisms are caught and thrown overboard* by commercial fishing operations. This deplorable wastage is equivalent to *one billion meals,* and routinely occurs in what the Federal government regards as, "America's best-managed Marine Fisheries." In 1998, nearly 100 massive catcher-processor ships were operating in areas of the Bering Sea and Aleutian Islands.

Conservation groups have recently called for immediate action to revise outdated 20-year-old policies that are "systematically destroying the world's largest and most productive ecosystems." For example, The Alaska Ocean Network Plan calls for more conservative fishing quotas, recognizing the relationships between a healthy marine environment and coastal communities, restrictions on bottom trawling, and establishing "Regeneration Zones."[46]

Size Matters

In the marine environment, size is a very important factor for positive genetic selection, so the time-honored tradition of taking "the Biggest and the Best" has undoubtedly had a disastrous effect on the marine biological gene pool.

Research has shown that larger fish are reproductively more prolific *by several orders of magnitude,* than smaller individuals of the same species. These larger individuals will also most likely have superior genes for growth and survival. For example, a legal-sized 17-inch red snapper produces about 44,000 eggs a year, while a larger 24-inch snapper can produce 9.3 million eggs. In other words, the larger fish can produce as many eggs as 211 of the smaller ones! The important thing to understand, is that size *does* matter! We simply need to revise our thinking, and "throw the big ones back," so they will have the chance to replenish our fisheries resources.

Fishing regulations have tended to follow the same old out-dated guidelines, simply due to a "lack of initiative for change." With world fisheries collapsing at an alarming rate, it is time to develop new fishing regulations which are compatible with our current world fisheries stocks, and thus appropriate for a sustain-able future.[47]

Other Size Benefits

A small snail has recently been identified as a possible cause in the recent destruction of Elkhorn corals in the Greater Caribbean region. This little snail just happens to be a favorite food of the Caribbean lobster, but the lobsters must reach a certain critical size before their mouth parts are strong enough to crush the snail's shell. "Legal-Sized" lobsters have not quite reached the size where they can crush the snails to eat them, but larger lobsters have been observed to gobble up as many as a dozen of the snails within just a few minutes.

Prior to the introduction of spear fishing in the 1950's, large lobsters (10 pounds, or more) were quite common in the tropical waters of the world. Since these large crustaceans were heavily armored, they were practically immune to attacks from natural enemies, once they had attained a certain critical size. They also undoubtedly played important roles in maintaining healthy and balanced reef ecosystems, by grazing on the reefs and surround-ing grass beds. However, these leviathans proved to be no match for spear fishermen, who found them easy prey, and sought them out in increasingly remote locations as lucrative "Prize Catches." As a result of this sudden assault by humans on the natural environment, large lobsters, and lobster populations nearly everywhere, have been diminished to a mere shadow of their former abundance.[48]

Coral Reefs: Our "Undersea Rainforests"

Because of their fantastic biological richness and diversity, coral reefs have sometimes been called the "Rainforests of the Sea." Scattered throughout 101 countries and territories of the Earth, coral reefs are critical global resources in terms of fisheries, coastal protection, and eco-tourism. Coral reefs function as "Biological Banks," since they provide habitats for up to two million marine plant and animal species. Scientists estimate that perhaps only ten percent of the unique species that form these reefs have even been described. Coral reefs represent valuable resources in terms of providing new biological compounds, the most noted medicinal compound perhaps being AZT, which is used in the treatment of AIDS. Coral reefs also function as "Genetic Banks," by protecting breeding populations of marine organisms, so they can continue to produce offspring to "seed" nearby fishing areas.

A recent satellite atlas of global coral reefs, compiled by the United Nations Environment Programme (UNEP) indicates that coral reefs cover *only one-half to one-tenth of the area formerly thought to exist.* The new atlas shows that coral reefs cover an ocean area of about 284,300 square kilometers (an area about the size of France), miniscule, when compared with the total ocean surface area, which covers about 70 percent of our planet.[49]

The world's coral reefs are subject to major ecological pressures which include: over-fishing, specimen collection of shells and marine organisms for aquariums and traditional medicine, marine pollution, siltation due to dredging and soil runoff, and coral bleaching due to global warming. Recent satellite studies indicate that *over ten percent* of the world's coral reefs have already been seriously damaged.[50] Like "Miner's Canaries," coral reefs function as sensitive "Bio-Indicators," since they reflect back the effects of environmental degradation and global weather extremes.

During recent years, coral reefs have taken on an increasingly important role as renewable resources for tropical islanders and coastal inhabitants. Since these reefs contribute a major share to the multi-billion-dollar ecotourism industry, a global system of coral reef parks could become financially self-supporting in the future. Coral reefs represent critical and sustainable resources for both present and future generations. They function to maintain biological diversity, and help sustain a natural balance within the marine ecosystems of the world.

In Indonesia,
where one-eighth of the world's Coral Reefs are located,
over 70 percent of the reefs are dead, or dying.
It is estimated that the Indonesian economy
loses between US $500,000 and $800,000 every year,
for every square mile of dead or damaged reef.

~ *The Futurist,* January-February, 2003, p. 42.

Rigs to Reefs

The state of Texas has had an artificial reef development program for over 50 years! Tires, discarded vehicles, and construction rubble have all been tried in various reef building projects. In the long run these materials proved inappropriate, since they tended to be broken up, or moved by major storms throughout the years.

The first attempts at artificial reef development occurred in the late 1970's, when 12 obsolete Liberty ships were sunk at five different locations in the Gulf of Mexico. Remarkably—over 30 years later—these sites are still biologically productive, and have even been enhanced with additional materials.

Declines in oil and gas prices in the 1980's, resulted in increasing numbers of drilling rigs being scrapped. To date, some 49 drilling rigs have been donated by participating oil and gas companies for use as artificial reefs.

The Artificial Reef Act of 1989 was designed to promote and enhance artificial reef construction for the state of Texas. By combining environmental expertise with recommendations from a citizen-based Artificial Reef Advisory Council, a set of guidelines for establishing and maintaining artificial reefs was put into action.[51]

Artificial Designer Reefs

"Reef Balls" are flat-bottomed, hollow concrete spheres, with a "Swiss cheese" pattern of holes. When placed on the ocean floor, they become artificial habitats for a variety of fishes and invertebrates. Reef balls are created by pouring concrete into different sized fiberglass molds. When the concrete hardens, a flexible bladder is inserted into the interior space through one of the openings. The bladder is then inflated, so the reef ball can easily be towed through the water to its designated site. When the air in the bladder is released, the reef ball sinks gently to the ocean floor. Reef balls ranging from 2.5 to over 5 feet in diameter can be combined into an aesthetically pleasing arrangement to create an artificial reef.

Reef balls serve as ideal underwater fish habitats, and also provide an excellent substrate for corals and encrusting organisms. Studies have shown that single reef ball can produce up to 400 lbs. (180 kg) of animal and plant biomass in a single year. Reef balls thus provide excellent homes for juvenile fishes and invertebrates, and have also been successfully utilized for live coral propagation.

To date, over half a million reef balls have been deployed worldwide. One ambitious new project, Reefball Asia, has launched a program to place a million reef balls in the coastal waters of Malaysia, for purposes of marine resource enhancement and environmental education.[52]

New Technologies for Coral Reef Restoration

A company called Applied Marine Technologies on the Caribbean Island of Dominica has developed a technology which has the potential to restore the world's coral reefs. AMT has established the first commercial coral farm in the Caribbean—and perhaps the world.

Founder/Director, Alan Lowe, has been involved in coral farming for over 15 years. His company consists of a dedicated team of engineers, divers, marine biologists, and coral specialists. The company's main objective is to cultivate coral under controlled conditions, and then replant it in reef areas that have been environmentally stressed, damaged, or destroyed.

Corals reproduce asexually, so a small piece of a "coral tree," transplanted to a damaged reef area can eventually grow into a large colony. Corals also reproduce sexually once or twice a year, producing up to half-a-million eggs, of which only a few would normally survive in the natural environment. Under controlled culture conditions, however, survival percentages can be increased dramatically to help re-seed damaged reef areas. AMT's research has shown that a few hundred pieces of cultured coral, transplanted to a damaged reef structure, can give a damaged reef a 165-year head start over a reef that goes through a natural recovery process.[53]

The Geothermal Aquaculture Research Foundation (GARF) is an Idaho-based non-profit organization, which was established to provide marine aquarium enthusiasts and scientists throughout the world with information on the culture and maintenance of coral reef organisms in saltwater aquarium environments.

A group of like-minded individuals, hundreds of miles removed from the marine environment, have used artificial sea salts and geothermally heated water to leapfrog the scientific establishment. They have developed successful culture techniques for many species of exotic corals and invertebrates, collected from the tropical oceans of the world. Besides successfully culturing many species of delicate stony corals, soft corals, and invertebrates,

GARF has created novel new techniques for sustaining the delicate balance in marine aquarium systems.

The organization has also developed a global network of saltwater aquarium enthusiasts and suppliers of living marine specimens, who are engaged in coral culture throughout the world. The group also supplies and educates the public about "Reef Janitors"—snails and hermit crabs that graze on algae and diatoms—thus helping to maintain a proper ecological balance in the closed-system aquariums.

GARF's unique techniques and culture systems represent an exciting new resource for restoring coral reef species that have either been damaged, or have disappeared entirely, due to the pressures of pollution, specimen collection, or global climate change. These new culture techniques also open up new possibilities for the commercial production of new pharmaceuticals—without disturbing the natural coral reef ecosystems of the world.[54]

Electric Reefs

According to Australia's Global Coral Reef Monitoring Network, the world's coral reefs are under a massive environmental assault. During the past few decades, it is estimated that over one-fourth of the world's coral reefs have died off. If present trends continue, scientists expect that at least another 25 percent are doomed to perish within the next 20 years.

Assaults on coral reefs come in many forms which include: physical destruction associated with specimen collection, anchor damage, and overfishing; and pollution from excess nutrients in the form of sewage and chemical pollution from cities and farmlands. Mounting evidence now suggests, however, that it is global warming that currently presents the greatest overriding threat to the world's coral reef ecosystems.

Coral polyps normally feed on tiny plankton and organic particles, suspended in the water column. Most healthy corals are also home to symbiotic algae called "Zooxanthellae," that manufacture food for the corals, and endow them with their variety of spectacular colors. Unfortunately, however, these symbiotic algae tend to be more sensitive than the corals, and are thus the first to be affected by rises in seawater temperatures. For example, an increase of only 1°C can cause a die-off of the algae—leaving the corals with a "bleached" appearance. Without the food energy which the zooxanthellae provide, the corals tend to become less resistant to disease and environmental stresses, and often die off following these bleaching events.

Mass-bleaching events have occurred on coral reef areas nearly everywhere in the tropical seas of the world. For example, in 1999, elevated temperatures killed off most of the corals in the Indian Ocean. A similar mass coral die-off occurred across the tropical South Pacific Ocean in 2002, with Australia's Great Barrier Reef being especially hard hit. According to scientist Ove Hoegh-Guildberg, Chairman of the United Nations Working Group on Coral Bleaching, "There is little doubt that current rates of warming in tropical seas will lead to longer and more intense bleaching events."[55] In addition, increased carbon dioxide concentrations may also prove detrimental to the health and survival of coral polyps.[56]

Since the 1950's artificial reefs have been traditionally formed from sunken ships, scrap metal, discarded tires, concrete, and construction debris. Although such artificial reefs have been quite successful in attracting fishes, free-swimming invertebrates, and encrusting organisms, they lack the ecological complexity of natural coral reefs.

Experimentation with "Electric Reefs" dates back to the mid 1990's, when Architect/Engineer Wolf Hilbertz, at the University of Texas in Austin, began a series of experiments in which he immersed electrodes in seawater, and passed electrical currents through the system in an effort to recreate the growth mechanism of shellfish and corals. When direct current was passed through the metal electrodes, an electrolytic reaction occurred. Later on, Hilbertz discovered that when direct current was passed through a basket-like meshwork of titanium covered with a thin layer of ruthenium oxide, the metal skeleton became encrusted with a form of calcareous limestone called "Aragonite," which could produce coatings up to 5cm thick in a year's time. From these experiments, Hilbertz reasoned that by generating electricity from on-site renewable resources such as solar or wind power, it would be possible to create cost-effective limestone "Reefs" of any shape and size. Theoretically, such manmade structures would require little maintenance, and would even have the capability to "heal themselves" when damaged. Since the artificially created limestone resembled natural reef substrate, it was thought that corals would colonize it readily.

Up until 1988, Hilbertz's discovery had been largely ignored by the scientific community until Caribbean coral expert, Dr. Thomas Goreau, contacted Hilbertz about his work. With basic funding support from the Global Coral Reef Alliance, the two scientists have spent over a decade working to construct their artificial "Biorock Reefs" in various locations throughout the tropical seas of the world.

A typical biorock reef costs about $1,000 to construct, and consists of a domed steel framework about 12 meters in diameter. The largest biorock reef so far constructed, one of seven in the Maldives, is 40 meters long and 8 meters wide. In Bali, 22 of these structures have already been installed. Power requirements for these reef structures are modest, since a structure of 100 square meters requires less than 300 watts to produce optimal growth conditions for the corals. Most of the biorock reef structures take their power from banks of solar cells, located either on shore, or on a floating raft, anchored adjacent to the reef structure.

So far, the electric reefs have attracted wide variety of fishes and other typical reef inhabitants. Researchers were also surprised to learn that the electrical current seems to provide the coral polyps with more energy for growth and reproduction. It is thus estimated that Coral Growth on biorock occurs three to four times faster than on ordinary natural substrate! Electric reef corals have also survived ecological stress events that killed most of the corals on nearby natural reefs. According to Goreau, "On our biorock structures, more than 80 percent of the corals not only survived, they flourished."

Goreau and Hilbertz envision expanding their "Artificial Coral Garden Concept" to grow electric reefs from tens to hundreds of meters in diameter, as a way to help protect the world's coral reefs from the threats of pollution and global warming. A global network of these "Coral Reef Arks," could ultimately provide a way to re-colonize large areas where natural coral reefs have previously died off.

According to Seferino Smith, operations manager of an eco-logical resort owned by the Ukupseni Community of Kuna Yala in Panama, "Coral nurseries are the only way to keep corals from extinction. At this resort, local groups are working together with the Global Coral Reef Alliance to build biorock nurseries to support both fisheries and tourism industries. We should multiply these nurseries all over the Kuna Reserve and around the world."

As an interesting sidelight to their reef-building technology Goreau and Hilbertz are also interested in the possibilities of ap-plying their biorock Construction Techniques to build foundations for seawalls and jetties. They estimate this approach could reduce the costs of conventional construction by as much as 90 percent, making it ideal for island nations, which are presently threatened by rising sea levels.[57]

Reef-Fish "Laundering"

The untold story of "Reef-Fish Laundering" and the related decimation of coral reef fishes in the tropical Pacific has recently become a major concern of conservationists and scientists the world over. Wealthy Asians, in keeping with the latest trends, consider it fashionable to dine in fancy restaurants which feature expensive, exotic, and beautiful specimens of live reef fishes— selected from restaurant aquariums, and prepared with great fan-fare to impress distinguished guests. It is estimated that Asian countries consume some 50,000 tons of these "Live Fish" every year!

The ornamental fish export trade for the global aquarium market also continues growing at an exponential rate. According to Cristina Balboa of the Washington, DC-based World Resources Institute, in 1971 only 200 species of ornamental fishes were imported into the United States. By the year 2000, this number increased to over 1,038 different species.

In Asia, live reef fishermen tend to have "unsavory" reputations, since they traditionally use cyanide to anaesthetize the fishes they capture. Recently, several companies, in a fisheries version of money laundering, have apparently attempted to hide the sources of their live fish catches in the Pacific. For example, in the year 2000, 45 percent of the total reef fish catch was reported as originating in Singapore—a country that has neither any significant coral reefs, nor fishery. There is little doubt that this increased live fish boom has been fueled by the increased demand for live fish in Hong Kong and Chinese markets, where the most desirable fishes include exotic groupers, coral trout, and hump head wrasse, which command as much as $40 a pound in Hong Kong.

Live reef fishermen are known to be active in Micronesia, Fiji, Kiribati, the Marshall Islands, Palau, and Papua, New Guinea. In order to overcome the costs of the great distances between fishing areas and markets, live reef fishermen who undertake longer voyages must fill their holds with catches in excess of 20,000 pounds. Since these live fish species are often large and relatively rare, the losses from the Pacific coral reef ecosystems tend to be disproportionately magnified. In 1997, for example, it is estimated that some 25 million fish, averaging two pounds each, were exported to Asia, with the majority of this harvest coming from the Pacific area. The grim reality of the situation is that, without adequate laws, monitoring, and supervision, there is little or no incentive for foreign fishing vessels to develop sustainable fishing practices in the Pacific Island regions.

Conservationists and fisheries scientists have recently launching a coordinated effort to regulate the ornamental reef fishery that exports fish for the global aquarium trade, to the point where it can conceivably become an ecologically sustainable fishery. Similar efforts are underway to regulate the live reef fish industry, which has a history characterized by ecological destruction and unsustainability. Some experts feel that the Live reef fish industry offers the potential for becoming a low-volume, high-value, ecologically sustainable industry—provided it can be properly managed and monitored.[58]

BioProspecting in the Ocean Depths

Deep-ocean ecosystems contain an untapped wealth of genetic resources, offering untold possibilities for the fields of science and medicine. Since deep sea organisms have adapted under extreme environmental pressures, they possess very unique structures and metabolic pathways, which make them valuable sources for new biological compounds. The recent discoveries of deep hydrothermal vent communities in the ocean depths have revolutionized the thinking of biologists and geologists on the basic parameters of life itself, as well as the mechanisms for sea floor mineral deposition. Recently, an entirely new biological kingdom, called the Archaea, was discovered. The unique chemosynthetic pathways of these bizarre organisms have extended our definition of life itself, and have also provided exciting new clues to the origins of primitive life forms on Earth.[59]

Basic Strategies for a Comprehensive Global Resource Management Plan:

1) A set of ecosystem-based management guidelines should be established, to provide for the designation, monitoring, and protection of terrestrial and oceanic wildlife preserves. Such protected nature reserves would serve as "Biological Banks" for preserving species diversity and ecological balances in a natural undisturbed state. These natural preserves would thus represent prototypical "Biological Reference Models"—baseline references for ecological preservation and restoration programs in areas where environmental damage has already occurred.

2) An effective global oceanic resources management plan should be developed and implemented. Such a global program would be designed to eliminate existing inefficient and destructive fishing practices such as bottom trawling, bi-catch wastage, and general over-exploitation of world fisheries resources. This program should also establish efficient, environmentally friendly fishing methods, and regional catch quotas that would fall within the sustainable limits of global ocean resources.

3) The global satellite sensing technologies of existing space agencies should be coordinated into a single comprehensive program for monitoring the global ecosystems relative to atmospheric, land and water pollution, desertification, forest fires, and global weather patterns. In short, the "Health" of the Planetary Biosphere.

4) A military "Environmental Task Force" should be created to train and coordinate international military programs to protect environmental resources, enforce environmental regulations, implement environmental clean-up projects, undertake environmental education programs on the efficient use of Earth's natural resources, and participate in the planning and implementation of large-scale planetary environmental restoration and enhancement programs.

METASTRATEGY III

ENLIGHTENED POPULATION MANAGEMENT

The Earth is a hothouse now.
Six billion members
of the human family and rising,
congregated together
on a spinning ball,
in stress, in ferment,
caught between what was
and what is yet to be.

~ Jean Houston, *Jump Time*

In his provocative book, *World War III*, Michael Tobias expresses the relationship between human technology and the Global Biosphere as follows: "Our planet is under siege, threatened by an ever more rapidly growing hoard of the most dangerous mammalian species of all time—*Homo sapiens.* Human beings have poisoned the earth, the water, and the air with a myriad of synthetic chemicals that are now present, often in highly dangerous levels, throughout the globe."[60]

Author Thom Hartmann, from a uniquely different perspective, writes, "…in 1987, humans became the most numerous species on Earth in terms of total biomass. Around 1990, we became the most numerous mammalian species on the planet, outnumbering even rats." Hartmann goes on to say there is more human biomass on the planet than of any other species. *Homo sapiens* already consumes nearly 50 percent of the world's fresh water resources, and over 40 percent of Earth's total "Net Primary Productivity" (i.e. the total food and energy available to all living things).[61]

Study Shows that People Take Up Most of the Planet

According to a report by scientists from the Wildlife Conservation Society (WCS) and Columbia University's International Earth
Science Information Network, humans take up approximately 83
percent of Earth's land surface with activities such as living, working, farming, mining, or fishing. People have taken over some 98
percent of the land that can be used to farm rice, wheat, and corn—
leaving relatively few areas of pristine wildlife preserves.

A new map, recently published on the Internet (http:// www.
ucs.org/humanfootprint), combines data from such sources as
population density, road and waterway access, electrical power
infrastructure, and areas taken up by cities and farmland. This map
graphically depicts the few remaining significant regions of natural
wildlife that still remain our planet. These wild areas include the
northern forests of Alaska, Canada, and Russia; the high plateaus of
Mongolia and Tibet; and much of the Amazon Basin.

According to WCS landscape ecologist Eric Sanderson, "The
map of the human footprint is a clear-eyed view of our influence
on the Earth. It provides a way to find opportunities to save wildlife and wild lands in pristine areas, and also to understand how
conservation in wilderness, countryside, suburbs, and cities are all
related."[62]

Cows Are Better Off Than Half the World's Humans!

For about half the world's population, the grim reality is this:
You would be better off being a Cow! According to UK author
Charlotte Denny, the average cow in Europe receives the equiva-
lent of US $2.20 per day in government subsidies and other aid.
Meanwhile, some 2.8 billion people in the poor countries of the
world are presently living on less than $2.00 a day!

When it comes to the matter of global inequality, the facts
present a truly grim picture! In the U.S., for example, the richest
25 million Americans receive an income which is equal to that of
nearly 2 billion people in Poor Countries. Even taking into account

the financial market declines of 2001-2, the assets of the world's three richest individuals exceeds *the combined incomes* of the world's least-developed countries.

The impact on global society of this extreme gap in life expectancy and quality of life is truly devastating! In Sierra Leon, for example, average life expectancy is age 37—a figure not experienced in the Western World since *before* the Industrial Revolution! With three out of every ten children dying before the age of five, infant mortality rates are now higher than they were in England in the 1820's!

Two hundred years ago, personal income in Great Britain—at the time the world's richest country—was three times greater than in Africa, which was then the world's poorest region. Today in Switzerland, the world's richest country, inhabitants have a per-capita income *nearly **80 times greater*** than in non-white South Africa—the world's poorest region.

An economist associated with the world bank has warned that, as television and movies continue to graphically depict the increasing disparity between the world's rich and poor, the day may come when the rich may have to isolate themselves in security-controlled enclaves, just to keep out the angry masses of the economically dispossessed.

Extreme poverty is a primary driving force for migration, as the best and brightest individuals from the world's poor countries immigrate to rich countries, seeking better educational and economic opportunities for themselves. For this reason, restrictions on immigration by first-world countries are being tightened to the point where the trafficking of illegal aliens has become even more lucrative than smuggling drugs.

At the United Nations Millennium Summit global leaders established a set of goals to cut global poverty by 50 percent within a 15-year period. Other objectives of the program included eradicating hunger, reducing the under-five mortality rate by two-thirds, and providing education for every primary school-age child. The costs for achieving these goals were estimated to be between 40 and 60 billion dollars—in addition to the current levels of financial aid

already in place. To put things into perspective, this amount represents *only about one-sixth of the total amount spent by western
nations on farm subsidies.*

In a recent assessment of progress toward the goals initially
set at the Millennium Summit, UN representatives issued a
warning that some 33 countries—representing 25 percent of the
global population—are expected to miss half the objectives
originally set at the Summit meetings. Most of these countries
are in Africa. Furthermore, if living standards continue rising at
their present rate, the UN experts estimated *it would take ISO
years to entirely rid the world of hunger.*

The UN experts also indicated that if Sub-Saharan Africa is to
get back on track, it will require extraordinary efforts on the part
of everyone concerned. Although this task is certainly daunting,
it is not impossible, since the proportion of those individuals
living in absolute poverty has already dropped from 24 percent
in 1990 to 20 percent in 2002, mainly due to rapid growth in East
Asia. In 1960, for example, South Korea and Senegal had per-
capita GDP's of $230. By the year 2000 South Korea's per-capita
GDP had risen to $8,910. Living standards in Senegal, however,
had barely improved at all, with a per-capita GDP of $260—an
increase of only $30 from the 1960 figure.

While 50 years ago the main export item from Korea was
human hair, today, the country is a hi-tech leader in silicon chip
manufacturing. Massive government support for this industry in
the 1970's paid off handsomely in the 1980's and 1990's, when
silicon chips became the building blocks for the present hi-tech
revolution.

In contrast to Asia, Africa is fettered by debt, conflict, and
unfavorable geography. In short, it faces an unequal struggle in the
current global arena. Whereas South Korea was allowed to protect
its developing electronics industry from competitors, Africa is
being required by the International Monetary Fund and World
Bank to open up its markets. The pattern that now appears to be
developing is that rich countries simply continue to advance—
pulling the development ladder up after them. As the accelerated

pace of globalization continues to encourage greater national independence, solving the global poverty issue becomes a political as well as a moral imperative.[63]

> **We have created
> this overcrowded world
> of overtaxed resources
> by consuming ancient sunlight,
> converting it into contemporary foods,
> and consuming these foods
> to create
> more human flesh.**
>
> ~ Thom Hartmann, 1998, *The Last Hours of Ancient Sunlight*

The Specters of Starvation and AIDS

Intense global population pressures have created yet another problem of vast proportions—the ominous "Specter of Starvation," which continues to wreak havoc, mainly in third-world countries. Each year, 12 million people will starve to death, and 30 million more will perish from hunger-related diseases. This is equivalent to 300 Boeing 747's with 400 passengers each, crashing every day of the year, with no survivors.[64]

Despite numerous advances in agricultural technology, one out of every five third-world citizen is still unable to get enough to eat. As global population continues to expand, both hunger and starvation persist in epidemic proportions in many areas of the world. These conditions will worsen as human population continues increasing.

Increased population pressures have coincided with the emergence of new and resurgent diseases. Over 50 years since the discovery of antibiotics, infectious diseases are again on the rise. Worldwide, over two billion people are infected with tuberculosis, cholera, malaria, diphtheria, sleeping sickness, and schistosomiasis. It is estimated that one out of every three people in the world now carries the TB bacterium! New pathogens have also recently appeared on the scene, that are resistant to all known forms of antibiotics.[65]

Added to this grim scenario are the devastating effects of AIDS and other sexually transmitted diseases. "Across Asia and Africa, AIDS has been referred to as the huge time bomb of the sex industry, because it is through prostitution that AIDS gains its widest spread. This epidemic is out of control in dozens of countries where almost no one is willing to act on it, or even talk about it."[66]

The Bitter Fruits of the American Sexual Revolution

Over 50 million Americans are infected with incurable Sexually Transmitted Diseases (STD's). As many as 12 million cases are reported each year, with 85 percent of them occurring in individuals between the ages of 5 and 29 years of age.

Human Papiloma Virus (HPV) is the most rampant Sexually Transmitted Disease in the United States, with over 70 different strains having been identified. Sexually Active Women under age 25 are apparently the most vulnerable, as some studies estimate the rate of infection to range from 30-40 percent in this group.

One out of every five American adults has Genital Herpes. Perhaps the most tragic aspect of this statistic, is that 90 percent of these infected individuals are not even aware they have the disease!

Chlamydia is now the most common cause of infertility in the U.S. This STD infects approximately four million people every year. Forty percent of all sexually active individuals are thought to be carriers of this disease, with seventy percent of these infected individuals being unaware of it.

According to the United States Department of Health and Human Services, teenagers are now contracting AIDS faster than any other age group.

~ Joseph Collison, *The Oxford Review,* May 2002.

New Birth Control Technologies

Recent advances in birth control technologies include seven-day transdermal contraceptive patches for women, and improved intrauterine devices. Norplant Long-Term Female Contraceptive Implants have also proven to be effective—requiring only one clinical visit a year. In a simple 10-minute procedure, six tiny rods, which give off small amounts of a Progesterone-like hormone, are inserted under the skin of the upper arm using a local anesthetic.

A different approach has been taken by Conceptus, Inc. of San Carlos, California. The company has created a new alternative to fallopian tubal ligation, which they have named "Essure." In a simple clinical procedure, slender coils of soft titanium-nickel alloy are inserted into the fallopian tubes, using local anesthetic. Once in place, the coils expand to fill the fallopian tube. As of September, 2000, 187 women had tried the device for 11 months, with no conceptions, or reports of pain, inflammation, or infection.[67] The device has already received FDA approval, which was expedited because of its potential benefit for couples seeking an alternative method to surgical sterilization. According to an FDA spokesperson, "It's different from other methods used for sterilization because it does not require an incision or general anesthesia."[68]

Other new approaches to birth control include: the FDA-approved Drug RU-485 (Milfeprex), for termination of pregnancy for up to seven weeks. A new male contraceptive pill, presently undergoing clinical trials in Scotland, China, South Africa, and Nigeria, has so-far proven 100 percent effective in suppressing sperm production. The pill, which is produced by the Dutch pharmaceutical company, Organon, is expected to be on the market within three or four years, pending final testing.[69]

Although effective birth control technologies do exist, development of family planning programs, and making these technologies conveniently available, are another matter. One prototypical example of how a single inspired individual can become an effective "Agent of Change," comes from ecologically stressed Indonesia. There, due to the heroic efforts of Dr. Firman Lubis, Director of the Yayasin Kusuma Buana NGO Women's Clinics, over 700,000 Indonesian women are now using Norplant Implants![70]

- **Over a million people die every year of Malaria.**

- **West Nile Virus, never seen prior to 1999, is spreading faster than ever in the U. S. By the end of the summer of 2002, eleven fatalities, and infections as far north as Maine and Minnesota had been reported.**

- **Only ten percent of the funding for Infectious Disease Research goes to these diseases that cause 90 percent of the deaths worldwide, because they occur in poor countries.**

- **Far more support is needed from Rich Nations, many of whom are now experiencing Tropical Diseases in their countries for the very first time.**

  ~ *New Scientist,* August 31, 2002, p. 3.

Eco-Hotel Brings Indigenous Women Together in Mexico

Although there is nothing unusual in Mexican hotels about women scrubbing floors, making beds, and cooking meals, the Hotel Taselotzin in Mexico's scenic eastern Sierra Mountains provides an interesting example of a Successful Women's Cooperative. Here, you can also find local women managing, bookkeeping, marketing, and tending organic gardens for their own internationally acclaimed eco-tourist resort.

The Women's Cooperative consists of over 200 Nahua Indian women from the state of Puebla, and is called Maseualisuame Mosenyolchikauani, which translates into "indigenous women working together." In addition to the resort, the group also manages a health food store, a greenhouse for medicinal plants, a traditional crafts center, and workshops, where the ladies produce recycled paper and biodegradable cleaning products.

Hotel Taselotzin sets an example both in terms of ecological design and function. The eco-hotel combines the ladies' indigenous perspective with knowledge gained from similar groups throughout the country. The hotel also has its own reforestation projects, ecological tours, a rainwater collection system, and a composting center.

According to Rufina Villa Hernandez, one of the eight Co-Op directors, "The organization has served women well in giving us a place in the family, in the organization, and in the community." The Co-Op began as an offshoot of an organization called Co-maletzin, formed in 1987, which was created to strengthen the Rural Women's Movement in Mexican Society—from both ecological and gender perspectives. Says National Director Andrea Garcia de la Rosa, "We choose to work with rural women because in their condition and position, they are the most marginalized. They carry more than their share of work, have little access to services and resources, and are undervalued." She continues, "We have learned from the rural women and men to value and connect with natural resources in an active and respectful way."[71]

Women and Eco-Justice in Iran

Victoria Jamal, Director of Environmental Research at the University of Tehran, along with several of her professional colleagues, has created Iran's first program of environmental law. Director Jamal was instrumental in founding the Women's Society Against Environmental Pollution, which has since become one of Iran's most active NGO's. The Society has a membership of over 2,500 members, and publishes its own bilingual journal.

Victoria Jamal is a prototypical pioneer in a country where women are normally discouraged from assuming leadership positions. The Women's Environmental Society has helped increase public awareness of the major ecological challenges which Iran is currently facing. These challenges include: severe air and water pollution, over-fishing and biological degradation of the Caspian Sea, and loss of endangered species such as the Persian cheetah.[72]

Population Boom (or Bust) for the Future?

Forget the "Population Bomb." There may be a "Baby Bust" in the future for Global Society! Recent statistics indicate that 60 countries, including North America, Europe, East Asia, and parts of the Caribbean, already have fertility rates that fall *below* population replacement levels. Other regions, including Thailand, Iran, Vietnam, and Sri Lanka, will reach this level within the next 20 years, should present trends continue.

In much of Southern and Eastern Europe, women already average only 1.3 children—a figure well below the 2.0 needed to sustain population levels. Recent studies by Joseph Chamie, head of the United Nations Population Division, indicate that fertility projections for most of the developing world will inevitably lead to eventual population declines, barring major increases in life expectancy.

Even though the women of Bangladesh are among the poorest and least-educated in the world, they are presently at the forefront of a revolution that has confounded population scientists. Astonishingly, in a single generation, they have cut their average birth rate by 50 percent! Today, these women average three children, a *drastic* decline from the average of six children for their mothers! These women—as well as hundreds of millions of others—could well be opening a new path of demographic decline for the human race. If these trends continue, within the next 50 years, four-fifths of the world's women are likely to settle for families of two children or less. Should this assumption prove correct, global population will truly begin to shrink. Preliminary results of the ongoing study indicate a surprising trend: Whether Catholic, Moslem, socialist, or capitalist, rich or poor, with governments in favor of contraception or not, women everywhere are choosing to have fewer children. UN population expert Chamie refers to the implications for global society as "Momentous."

The U.N. Study Results present a startling contrast to the dire predictions of exponential population growth put forth in the 1970's. Even at that time, however, there were some indications that the impending population bomb had been defused. Since the early 1950's, when the average global woman had a family of five, fertility rates have continued to decline. Astonishingly, this number has already been reduced from 5 to 2.7!

Demographic experts had theorized that the rest of the world would eventually follow the path taken by Europe and North America; moving from a state of "Population Explosion," to one of "Population Stability." In general, populations have tended to spike upward, as improved nutrition, sanitation, and medical services have reduced mortality rates. Following this initial increase, however, these same populations have tended to stabilize as mothers decide it is better to have fewer children.

Until rather recently, scientists believed that the global population, which quadrupled to a figure of six billion by the end of the 20th century, would stabilize to between 10 and 12 billion by the end of the 21st century. The recent UN projections now suggest that, after peaking at 9 billion by the year 2050, world population will begin to drop. This type of decline is already evident in many countries of the world. For example, by the year 2050 Japan's population is expected to drop by 14 percent, Italy by 25 percent, and Russia by 30 percent. From the perspective of Australian demographics expert Jack Caldwell, "There no longer seems to be any barrier to most countries falling below replacement levels." Jack Chamie's projections from the UN study suggest that by the end of the 21st century, there could actually be *fewer* people on the planet, than there were at the beginning.

Two major factors have recently changed demographic thinking: 1) It has become apparent that people are *not* necessarily obsessed with the idea of having at least two children, and 2) The concept that women have to reach a certain level of education and prosperity before they will voluntarily reduce the numbers of children they have, is simply untrue. The past two decades have borne witness to the fact that ever-poorer countries have joined the trend toward having smaller families. Bangladesh, among the half-dozen poorest nations outside Africa, is a prime example. Equally unexpected, is the fact that most of the decline in fertility rates has occurred among women who are illiterate. Surprising too, is the fact that many of the countries with the most rapid birth rate decline have governments that are opposed to state-sponsored family planning. In Brazil, for example, the fertility rate has fallen from over six children per family a few decades ago, to just over two children today. In Iran, the fertility rate has dropped by two-thirds in less than twenty years. It thus seems that women everywhere, regardless of their level of education or economic status, are seizing the opportunity for better lives—which *does not* necessarily involve having large numbers of children.

Griffith Feeney, of the East-West Center in Hawaii, explains the underlying reasons for this transformed mindset as follows: "The enormous time, energy, and emotion women used to spend on bearing and raising children, most of whom died before reaching adulthood, can now be spent on other things." Says Tom Dyson, of the London School of Economics, "Getting married and having children are simply not as important as they used to be."

Countries which are exceptions to the declining fertility rates include Argentina, Israel, and Malaysia, which have sustained fertility levels at about three children per family. In the United States, because of the massive influx of immigrants, fertility rates have not dropped as dramatically; and in the poor African and fundamentalist Islamic countries, women still continue to produce six children or more. Moreover, the widespread AIDS Epidemic—especially in the poorer regions of Africa—is distorting fertility statistics, since the relatively high death rates of young individuals tend to push the true fertility replacement figures to over three children per family.

The new demographic patterns which are presently emerging worldwide suggest possibilities for sweeping social changes in the future, with migrant workers becoming a formidable force in countries like the United States. As childbirth continues to decline, the global population of elderly people will continue to grow about twice as fast as the general population rate. This rapid growth of the "Elder-Population" will put increased pressures on financial and healthcare services.

In 2003, the average age of a global citizen is 28. By 2050 this figure is expected to increase to 40. Opinions differ on the eventual outcome of this major shift in the average age of the global population. Some, like Ben Wattenberg, of the American Enterprise Institute, depict the ageing global population as "The Real Population Bomb." Others, like author Theodore Roszak of Cal State University, suggest that this ageing of the general population will tend to create an older, wiser, and gentler world.

Over the past three decades, we have come to the stark realization that overpopulation is a major threat to the continued well-being of both the human race and the Global Biosphere. A declining world population would most certainly help resolve many of the ecological and socioeconomic problems that are closely related to overpopulation. It is important to remember, however, that a major downward shift in global population will inevitably result in correspondingly dramatic shifts in the attitudes and social structure of global society as a whole.[73]

A Plan for Achieving Sustainable U.S. Population

If America is to maintain its position of global leadership, it must work toward achieving a sustainable population, a population which exists in a state of synergistic balance with the natural environment. We must achieve this goal in order to *halt* and *reverse* the present trends toward environmental destruction and degradation. We need to develop a sustainable global economy, which is both practical and realistic for present and future generations. To achieve this goal, we must first address the major root cause of our environmental and economic problems— overpopulation!

Based on recent U.S. Fertility and Immigration Rates the U.S. Population (presently over 288 million) is projected to exceed 400 million by the year 2050—with no end in sight! Unless we can effectively reverse the present rate of population increase in the U.S., all efforts to save our already fragile environment are likely to be futile! The ultimate objective should be for America to establish

a prototypical example for the rest of global society, by first stabilizing our own population at a reasonable, sustainable level;
a level *far lower* than what exists today. This quest for a
stabile, sustainable society *is absolutely critical* to the health,
happiness, and well-being of everyone, since the impacts of
human activities on the environment are now tremendously
magnified—simply because of the increasing masses of humans
attempting to occupy the finite living space on this Planet.

According to guidelines established by the non-profit organization, Negative Population Growth, Inc. (NPG), this goal
might best be achieved by reducing the U.S. population from its
present 288 million plus, to a level of 125 to 150 million over the
next three to four generations. Such a reduction would effectively
stabilize U.S. population at a level equivalent to America in the
1940's. This critical objective could be achieved by taking two
simple actions, which would result in a positive future lifestyle
scenario for people of all ages. First, *all* illegal immigration should
be halted. Second, legal immigration quotas should be limited to
100,000 or less per year. These two actions alone could halt the
main causes of population growth in the U.S., as well as in other
first-world nations.

Another major strategy for reducing the *rate* of population
increase would be to lower the fertility rate (the average number
of children per female) from the present level of 2.1, to 1.5 or
less, and then maintain this reduced level for several decades.
Put simply, if almost all women of child-bearing age had *no
more than two children each,* the U.S. fertility rate would
level off at about 1.5, since many women—especially those
with professional careers—are deciding to remain "childless
by choice," or simply having only one child. NPI feels that
promoting a reasonable ideal of "two-children-per-family" as
the social norm is the key to lowering the U.S. fertility rate,
although they also feel that non-coercive financial incentives
may also be required to help meet this goal.

The financial incentives proposed by NPG are designed to
shift general public attitudes away from "Entitlement," and toward

"Self-Responsibility." 1) Eliminate the present Federal tax exemption for dependent children born after a specific date, 2) Provide Federal income tax credits *only* to parents who have no more than two children, and 3) Provide an annual cash grant to assist low-income parents who pay little or no income tax, and have two children or less. Parents with more than two children would not be eligible.

Implementation of these guidelines would effectively change people's thinking, and reduce *both* immigration and fertility rates, so America could embark upon a path toward a sustainable population of between 125 and 150 million people—where happy, healthy, and prosperous lifestyles could be enjoyed by nearly everyone. Through an enlightened global education program, which would teach children at an early age to understand the values and benefits of sensible population management, it should be possible to halt, and then reverse global population growth; eventually correcting the financial and ecological inequities which exist nearly everywhere in the world today. For this type of enlightened social program to be successful, political and religious leaders would need to find the courage to implement the necessary and sensible co-evolutionary shifts to create new political, cultural, and religious guidelines, which would lead global society away from the socially inappropriate traditions of the past, to accept new concepts which are socially relevant and environmentally sustainable for both global society and the global environment.[74]

Carlos Hernandez and Rashmi Mayur in their book, *Pedagogy of the Earth,* stress the importance of abandoning the archaic and ecologically inappropriate concept of male domination in matters of sexual procreation. They feel that society needs to take a much more enlightened view of family management, where men and women *both* share in the decisions and responsibilities associated with creating and raising children, with the women having the final say, since biologically and socially it is they who bear the major share of childcare responsibility—especially in third-world societies. They believe that, "Every society must challenge the male domination and control of fertility." They go on to say, "In fact, all the family planning programs in every society should be in charge of women, who have a key role to play in social development, environmental protection and population sustainability."[75]

Basic Strategies for an Enlightened Global Population Management Plan:

1) A coordinated global program should be established which teaches the distinction between the concepts of sex and reproduction. This would involve the development and coordination of sensible programs for sex education and responsible family planning, combined with effective applications of birth control technologies.

2) A global shift in thinking must take place in order to bring cultural and religious family planning attitudes into a 21st century framework; a framework which is compatible with a sustainable future scenario for all humanity. Such a global attitudinal shift is necessary to reduce population pressures, poverty, and disease. This will require a coordinated effort between government, educational, religious, and corporate leaders to support sustainable population management strategies, with the ultimate objective of eliminating poverty and disease everywhere on the Planet. By shifting the social focus away from outdated customs which encourage creating many children, smaller families would be encouraged. This will conserve valuable time, energy, and expense for the parents, and will allow the women of the world to achieve a new levels of creativity and happiness. In the future, it should become more socially desirable for career mothers (and fathers) who wish to bypass pregnancy and biological birth, to adopt orphaned or socially disadvantaged children. These children would then have the chance for a new life with responsible parents, who have the desire and financial means to care for them, providing them with opportunities for a healthy lifestyle, a positive family environment, and unlimited possibilities for a bright future.

3) New programs should be established to re-direct the biological drive for genetic reproduction into alternative creative

pathways. Thus, for women all over the world, the creative drive for reproduction (A biological holdover from ancient times) can, if so desired, be transposed into new levels of creativity and leadership for women in areas of business, science, education, the arts, and politics.

4) Empowerment of women in family management should be encouraged, and free birth control technology should be made available for *both* men and women. To be truly effective, population management must become an integral part of sustainable social thinking at the grass-roots levels of societies everywhere.

5) Research and development of effective, inexpensive long-term contraceptive methods should be intensively funded and coordinated on a worldwide basis. Family management and sustainable ecology are intimately connected—as both are critical for the survival and well-being of all life on this Planet.

6) Rich nations must work together bring reasonable living standards to people everywhere on Earth, as starving people, without hope for even the basic necessities of life, can *never* be concerned with anything but their own survival.

METASTRATEGY IV

A GLOBAL SHIFT TO CLEAN, RENEWABLE ENERGY TECHNOLOGIES

**This country should lead the world
in a historic transition
from a society that runs on Carbon
to a society that runs on Light.
Such a Solar Hydrogen Energy Economy
would truly be the mark of an Advanced Civilization.**

~ Michael J. Osborne—2001, *Lightland:
Climate Change & the Human Potential*, p. 13.

Pollution of the Global Atmospheric Commons

"Rising industrialization in Asia discharges tons of contaminants into the winds traveling across the Pacific Ocean. Aerosols, which kill crops and spread illness in Asia, now also pollute the waters of America, and could dramatically alter global climate."[76] In April of 2001, an enormous dust cloud, which originated in Mongolia, was photographed by satellites as it moved across the Pacific Ocean, and over the West Coast of the United States (http://www. lakepowell.net/impact2001.htm). This massive cloud extended for 1200 miles, creating a whitish haze from Phoenix, Arizona to Alberta, Canada, where normal levels of airborne particulate matter quadrupled.[77]

Due to the increased demands for global energy, atmospheric carbon dioxide concentrations have increased by more than 50 percent over the past 50 years.[78] The majority of this pollution is caused by coal-fired industry and power plants. Approximately 28 percent of Earth's coal resources are located in the United States. Although coal represents approximately 90 percent of America's energy reserves, it represents only about 30 percent of its energy consumption. In China 75 percent of all energy is generated from coal. Acid rain from Chinese sulfur dioxide emissions has already caused extensive damage to human health and the environment, as these toxic emissions are responsible for over 50 percent of the acid rain in Japan, and 80 percent of the acid rain in Korea.[79]

Soot: A Major Cause of Global Warming

Recent studies indicate that soot just may be the second greatest contributor to global warming—just behind carbon dioxide as a major global atmospheric pollutant. Soot (black carbon) is thought to be responsible for 15-30 percent of global warming, yet it is rarely factored into current discussions of global climate change. In fact, only a handful of scientific studies have addressed the effects of soot on global warming.

Soot is composed mainly of elemental carbon, with ninety percent of it being generated from fossil fuels such as coal, jet fuel, natural gas, wood, and forest fires (Modern gasoline engines emit virtually no soot). Scientists have recently discovered that soot particles combine with other particles in the atmosphere, and that atmospheric mixtures containing black carbon absorb *twice as much heat* as pure carbon particles. The intense sunlight in the tropical areas of the world heats the soot in polluted air, which has the effect of burning off the top layers of cumulus clouds, often hundreds of miles downwind from pollution sources. This effectively reduces the heat-shielding effects of the cloud cover, and causes an increase in ocean surface temperatures.

In many countries, policies have been put in place based on the misguided assumption that since diesel vehicles get better mileage, then diesel fuel is better for the environment. Currently, about one-forth of all cars, trucks, busses, and tractors run on diesel fuel. Global restrictions on soot emissions could thus be an effective way to counteract global warming.[80]

According to Stanford University Engineer, Mark Jacobson, "If you want to control global warming, the first thing you go after is the soot." Jacobson claims that soot from burning diesel fuel and wood has a much more significant impact relative to its mass, than do the other major greenhouse gases, carbon dioxide and methane. He feels that, by eliminating all fossil fuel soot (some five million tons per year on a global basis), we could conceivably cut global warming effects by as much as 40 percent in a space of only three to five years!

Whereas soot particles warm the air by absorbing sunlight and radiating it back into the atmosphere, greenhouse gases create warming by absorbing heat, and radiating it back into the environment. Based on this logic, removing soot from the atmosphere would have *a more significant effect* on global temperatures, since Soot remains in the atmosphere *for a much shorter time.* Most existing climate change models *do not take Soot into account at all;* neither was soot factored into the 1997 Kyoto Protocol.[81]

The World's Dirtiest Fuel

The ships that ply the world's oceans and waterways carry 90 percent (by weight) of all the world's trade goods. Large ships (including cargo freighters and ocean liners) represent one of the major pollution sources on the planet. Despite their significant impact on the Global Biosphere, these "Seagoing Pollution Generators" are still largely ignored and mostly unregulated. What these vessels all have in common, is that they burn bunker oil—the dirtiest and cheapest fuel available!

In major American ports, such as New York, Baltimore, and Los Angeles, as well as in major ports around the world, ships account for as much as 17 percent of all nitrogen emissions during the summer ozone peaks! Within this context, a single ship creates more air pollution than 2,000 diesel trucks. Such unregulated toxic emissions make ships a major factor in the creation of smog and acid rain, which distort the global weather patterns.

In February of 2001, the Washington DC based NGO, The Earth Justice Legal Defense Fund, filed a lawsuit designed to force the U.S. Environmental Protection Agency to set tough new emissions standards for ocean-going ships.[82]

An Enlightened Energy Conservation Idea

In India, a one-million-dollar factory could produce enough fluorescent light bulbs to save the Country 204 billion dollars annually, by not having to build additional power plants.

~ Ruth L. Sivard, "World Military and Social Expenditures 1991," World Priorities Publications, Washington, DC, p. 41.

Asia's "Big Brown Cloud"

In August of 2002 a two-mile thick hazy brown cloud blanketed Southern Asia, disrupting seasonal monsoon weather patterns, causing agricultural damage, and putting at risk the health of hundreds of thousands of people in the region. According to U. S. Scientists the pollutants responsible for this brown haze could eventually result in "several hundreds of thousands" of premature deaths from respiratory disease—especially in countries such as Afghanistan, Bangladesh Bhutan, India, the Maldives, Nepal, Pakistan, and Sri Lanka.

According to Klaus Toepfer, Executive Director for the UN Environmental Programme, "... initial findings clearly indicate that this growing cocktail of soot, particles, aerosols and other pollutants are becoming a major environmental hazard for Asia." He went on to say, "The haze is the result of forest fires, the burning of agricultural wastes, dramatic increases in the burning of fossil fuels in vehicles, industries and power stations, and emissions from thousands of inefficient cookers burning wood, cow dung and other biofuels."

Current studies in the Asian Region indicate that fatality levels tend to rise dramatically with increased levels of atmospheric pollutants. Readings from seven major Indian cities have shown that, in the early 1990's air pollution was responsible for 24,000 premature deaths annually. By the mid 1990's this figure had risen to 37,000. The significance of such massive-scale pollution to the global atmospheric envelope was underscored by Executive Director Toepfer, who warned, "There are also global implications, because a pollution parcel like this, which stretches three kilometers high, can travel halfway round the globe in a week." For environmental scientists, the overriding concern is that both regional and global impacts of pollution-generated haze will intensify dramatically as the Asian population continues increasing to an estimated five billion people.

UNEP Scientists are calling for an Environmental Action Plan to address pollution threats in the Greater Asian Region, since the type of pollution blanket generated there can effectively reduce the amount of solar energy reaching Earth's surface by 10 to 15 percent. Concurrently, the heat-absorbing properties of this haze are thought to be significantly warming the lower layers of the atmosphere. The combination of surface cooling and lower atmospheric heating appears to be altering winter monsoonal patterns; causing dramatic decreases in rainfall in the northwestern regions of Asia, and increased rainfall along the eastern coast.

As the UNEP presented its findings at the 2002 Summit in Johannesburg, Executive Director Toepfer declared, "We have the initial findings, and technical and financial resources available. Let's now develop the science and find the political and moral will to achieve this for the sake of Asia, and for the sake of the world."[83]

California Dream or Toxic Nightmare?

According to a recent report by the Washington, DC based National Environmental Trust, a two-week-old baby in urban Los Angeles has been exposed to more toxic air pollution, than what the U. S. Government considers acceptable as a cancer risk for a person over an entire lifetime! The report also indicated that even if a young child moved away from the state of California, or if the air was cleaned up by the time the child reached adulthood, "the potential (cancer) risk a child rapidly accumulates in California from simply breathing will not go away."

California (the nation's smoggiest state) already has an adult cancer risk factor that is hundreds of times *above* acceptable levels for pollutants set by the Environmental Protection Agency. The report emphasized the fact that children were much more vulnerable to pollutants than adults because—in relative terms—they breathe more air, drink more water, eat more food, and play outdoors more than adults. The report went on to say, "A baby born in California will be exposed to such high levels of toxic air contaminants that the child will exceed the Environmental Protection Agency's (EPA) lifetime acceptable exposure level for cancer at a very early age, and will exceed the lifetime acceptable exposure level by many multiples by age 18."

The report claims that diesel exhaust from trucks, cars, busses, farm, and construction equipment is still the leading cause of pollution. It also cited chemicals emitted from dry cleaners and factories, plus pesticides, adhesives, and lubricant oils. The report recommended that state and local governments shift to using alternative clean fuels and technologies, replace diesel buses and municipal vehicles with cleaner alternative fuel technologies, and enforce existing fuel emission regulations.[84]

Oil from Coal—Free!

Prior to the year 1860 and the beginning of commercial petroleum production, 50 industrial plants were active in extracting Oil from Coal. In the 1920's, however, Lewis Karrick, a shale technologist with the U. S. Bureau of Mines, refined and perfected a process called Low-Temperature Carbonization (LTC). The Karrick LTC Process heated coal to a temperature of 800°F in the absence of oxygen; thus distilling oil from the coal, rather than burning it with oxygen. From a ton of coal, the Karrick Process yielded a barrel of Oil, 3,000 cu. ft. of rich fuel gas, and 1,500 pounds of smokeless coke. The economics of the process were such that the oil was produced—as a free bonus! The coke could be used in coal-fired boilers, or for coking coal in steel smelters, while coal gas produced by the process yielded more BTU's than natural gas, and produced only carbon dioxide as an exhaust gas.

When Karrick and his associates demonstrated that they could produce oil from coal cheaper than the oil wells (in addition to the bonus quantities of gas and coke), his pilot plants in Colorado were dismantled, and the process was suppressed by groups with vested interests in the oil and coal industries. Although Karrick did not actually invent the LTC process, he was instrumental in perfecting it. He was issued 16 U.S. patents for the process. Since these patents have all expired, they are presently available for use by anyone in the world.[85]

A New Technology for Combining Solar Energy and Wind Power

Australian company EnviroMission Ltd. has come up with a new way to harness the sun's energy to produce reliable, pollution-free electrical power. The company plans to construct a 3,300-foot-tall concrete chimney tower in the Australian outback. When completed, this will be the tallest manmade structure in the world, over twice the height of Malaysia's Petronas Towers—presently the world's tallest building.

This mammoth-scale power plant will consist of the 3,300-foot solar chimney, with a solar greenhouse at its base. The base of the tower will expand outward, much like the bell of a trumpet, to the width of a conventional football field. The chimney tower will be positioned over the center of a vast circular glass-roofed greenhouse, which will encompass an area of approximately 7.5 square miles. The solar tower complex, estimated to cost about US $563 million dollars, has been designed to generate 200 megawatts of clean electrical power.

Despite the massive scale of the solar tower complex, the technology involved is basically simple. The concept is unique in that it represents the hybridization of two different alternative energy modalities (solar energy and wind power). Sunlight heats the air under the glass greenhouse roof, which slopes upward from three feet at the outside perimeter, to 82 feet at the tower's base. The powerful updraft, created by the heated air rising up through the tower, spins an array of 32 turbines. In order to keep the generating plant working throughout the evening hours, the greenhouse will contain a network of water-filled tubes. Water in these tubes would store heat during the daylight hours. This stored heat would continue to heat incoming air during the evening hours, and thus sustain the updraft of air in the central chimney.

After dealing with the initial public skepticism for their mammoth project, EnviroMission CEO, Roger Davey, stated, "We have gotten to the point where it's not *if* it can be built, but *when* it can be built." The Project has the support of both the Australian and New South Wales Governments, who have designated the program "A Project of National Significance." According to Australia's Electricity Supply Association, 7 billion U.S. dollars of investment in electricity will be required over the next twenty years, in order to cover Australia's growing energy demands. The new Solar Tower Plant will generate up to 200 megawatts of electricity—enough to power 200,000 homes. This translates into about 650 gigawatt hours being contributed towards Australia's mandated renewable energy target, which requires that electricity producers supply 9,500 gigawatt hours by the year 2010. From an environmental perspective, the company expects that its MegaProject will replace approximately 700,000 tons of greenhouse gases yearly (which might have otherwise been emitted from conventional fossil-fuel power plants).

EnviroMission has signed agreements to conduct commercial feasibility studies with California-based Energen Corporation, and an Australian Company, Leighton Holdings Ltd. EnviroMission hopes to begin construction for the Project before the end of 2003, and to have the power plant fully operational by the beginning of 2006 (http://www.enviromission.com.au). An excellent conceptual video on the solar tower concept is available on the

following website: http://www.dailymotion.com/video/x11i2_
solar-tower-3d-animation_3d).

The Solar Tower Project was recently included in *Time
Magazine's* "Best Inventions of the Year." Says CEO Davey, "We
have proved that it does work and that it can be built, but what we
have got to get a handle on is the cost, and we are working very
strongly through that now."

The Solar Tower is a technically updated version of an earlier
design called "the Solar Chimney," originally conceived by German
structural engineer, Sclaich Bergerman, who in 1982 constructed a
656-foot-tall Demonstration Power Plant in Manzanares, Spain. This
50-Kilowatt Plant successfully produced electrical power for seven
years, but was shut down once the feasibility of the technology had
been demonstrated. Bergerman now continues his research with
EnviroMission in Australia. The new solar tower generating plant
will be constructed in the remote Buronga District of New South
Wales. The company expects that the high capital construction costs
of the project will be offset by low maintenance costs, the 100-year
practical operating life of the system, and the complete absence of
environmental pollution.[86, 87]

New Geothermal Technology

Texas scientist, Doyle Brewington, has developed a new device
for tapping Earth's vast geothermal energy resources. If this new
device works according to plan, it has the potential to revolutionize
the field of geothermal energy production. Brewington's so-called
"Power Tube" is designed to tap into Earth's subsurface heat reservoir far more economically than other existing geothermal-electric
power systems. The new system consists of four modules, which,
when put together, resemble a giant syringe.

The lowest module is a heat exchanger, where the internal fluid
(an iso-pentane/iso-hexane mixture) is heated to a vapor state. The
heated vapor passes on to the second module—an axial turbine
that turns a generator at the top of the stack. The fluid then passes
through a unique sound-powered condenser, after which the cooled

liquid flows back down through the outside layer of the tube, where the cycled is repeated.

The first prototype power tube, currently under development, is 29 inches in diameter and 85 feet long. It is designed to generate one megawatt of electrical power, at a working temperature of 215°F. Two other prototype units (also under construction) are destined for installation in Costa Rica and Hawaii. Larger ten-megawatt versions of the power tube device will be 185 feet long, and 42 inches in diameter. Power tube modules are designed to be transported separately, and can be conveniently stowed inside a C-130 cargo plane.[88]

A Geothermal-Powered UK?

An innovative mega-scale renewable energy project is being planned in the U.K., where a group of scientists have come up with a plan to develop a massive underground geothermal power plant, which could conceivably provide power for all of Great Britain. The $115 billion dollar power station would be located six miles underground, beneath Cornwall in Southwestern England. The installation is designed to produce 10 gigawatts of Electrical Power—enough for the entire country! By applying a process called "Hot Dry-Rock Technology," water is pumped into the Earth's core, where it is transformed into steam at 750°F (400°C), which can then be converted via turbines into abundant, non-polluting electrical power.

The initial three- to five-year research phase of this mammoth-scale project is estimated to cost US $86 million, and would take 25 to 30 years to complete. The newly created mining and heat consortium is convinced that, despite the high initial monetary investment, the project is both practical, and financially viable.[89]

Passamaquoddy Tidal Power

President John F. Kennedy believed that harnessing the tidal energy of Passamaquoddy Bay, near the eastern U.S./Canadian border, would be, "one of the most astonishing and beneficial enterprises undertaken by the people of the United States." President Kennedy envisioned transforming the bay's huge tidal fluctuations of 70 billion cubic feet of water into 50,000 kilowatts of electrical power. This dream that almost died with JFK is apparently being reborn in other areas of the world.

In May of 2001, the British Parliament issued the following statement: "The world can no longer afford to neglect the massive potential of wave and tidal energy." In June 2001, the state of Washington passed a bill requiring in-state utilities to begin providing Tidal Power and other forms of renewable energy as early as January 1, 2002.[90]

A New WavePower Turbine Technology

The Natural Resource Council of Canada has developed a relatively cheap, efficient, and non-polluting technology to convert the energy of ocean currents into electricity.

The Davis Hydro Turbine, developed by Canadian aerospace and hydrodynamic engineer Barry Davis, is linked directly to two other remarkable engineering achievements which preceded its development, the Avro Supersonic Jet (developed by the Canadian Government), and the D'Havilland Bras D'Or 40 Naval Destroyer, (produced by the Canadian Navy). The Davis Hydro Turbine might best be described as, "an Ultra-Efficient Underwater Windmill." The system is designed to produce from 200 to 8,000 megawatts of electrical power from the ocean tides.

A Canadian company called Blue Energy Power Systems is presently working on what has been designated as, "The World's Largest Renewable Energy Project"—a massive four-kilometer "Tidal Fence," spanning the San Bernardino Strait in the Philippines. The Project incorporates 276 generators, each of which is designed to produce 2.2 gigawatts of electrical power. Excess power will be utilized to run a desalination plant, and to produce clean hydrogen fuel. A bridge will be built across the string of caissons that supports the turbines.[91] (http://www.bluenergy.com)

Ring-Tailed Terrorist Attack

**In June of 2002, the unauthorized activities
of a Mischievous Raccoon,
were blamed for shorting out
Mississippi's Grand Gulf Nuclear Power Plant.
The 173-megawatt power plant
was knocked offline for over 48 hours!**

~ *Earth Island Journal,* Winter, 2002-2003, p. 12.

Dam-Free Hydropower

Northwestern University Professor, Alexander Gorlov, has developed a new technology for producing electrical power from flowing water, which does not require a dam. The "Gorlov Turbine" is designed to generate "Dam-Free Electrical Power" in streams and tidal channels. The horizontal turbine functions to convert a stream of kinetic energy into electricity at 35 percent efficiency (compared with about 20 percent efficiency for conventional turbines).

Four Gorlov Turbines have already been tested in Cape Cod, Massachusetts, and have functioned well. Another five-kilowatt system, installed in a pool in Vinalhaven, Maine, was able to generate sufficient electricity for a 15-room motel. Professor Gorlov envisions a wide range of potential applications for his innovative, environmentally friendly small-scale technology.

About ten percent of America's electrical power is presently generated by conventional hydropower. Gorlov estimates, however, that over 90 per cent of the energy in moving water is located in areas which are unsuitable for building dams. The new Helical Turbine thus provides a unique small-scale solution for producing ecologically appropriate, non-polluting hydropower, while making more efficient use of our finite natural resources.[92]

More Efficient Electric Motors:
A Key to Global Energy Conservation

Nearly 25 percent of the electricity in the U. S. is consumed by electric motors, which run intermittently, or 24 hours a day in homes, businesses and industrial centers everywhere. According to energy experts, even a one-percent increase in efficiency would result in millions of dollars in savings.

According to Ted Jones of the Boston-based, non-profit Consortium for Energy Efficiency, "Manufacturers tend not to care what it costs to operate a motor so long as it's productive." The major concern of industrial managers is reliability and the short-term objective of getting a motor back on line as soon as possible, instead of considering its efficiency, which would save money in the long run. This creates a situation where prices for new motors tend to be driven by the price of the equipment, rather than by its energy efficiency. The main problem with this kind of short-sighted thinking is that the actual cost of a large industrial motor typically amounts to *only about three percent* of its total lifetime operational cost—which can run up to 20 years or more. According to Jones, this means that essentially, what you pay for energy over time amounts to approximately 97 percent of your total cost! When he explains to industry executives that the cost of the original motor includes energy costs for the motor's lifetime, their response is something to the effect of, "I didn't know I signed up to spend $80,000 for a $2,000 motor."

The Consortium for Energy Efficiency, with support from the U. S. Department of Energy and similar groups, is attempting to raise the level of awareness of top executives and decision makers in a national campaign called "Motor Decisions Matter." The group hopes to demonstrate how a little extra initial investment in a more efficient motor—one certified by the National Electrical Manufacturer Association—can yield returns in energy savings, even within one to two years. Several major corporations such as Weyerhaeuser, 3M, and Disney World in Florida, have been receptive to the new standards. They have thus purchased the more-efficient "NEMA Premium Motors," which are rated for their high energy efficiency.

The efficiency of electric motors has been improving gradually since 1992, when Federal Energy Efficiency Standards were established for industrial electric motors. Says Rob Boteler of Emerson Electric Motors in St. Louis, "Electric motors are everywhere. Just look around your office. You can't find something that hasn't been touched in some way by an electric motor during manufacturing." Boteler feels that industry needs to view electric motors as a production cost, similar to the way labor and materials are factored in. He comments, "If you look at motors as a cost of doing business, you change your entire approach and improve efficiency."

The Efficiency Campaign is presently being expanded to include small motors of less than one horsepower, since they presently are not federally regulated—and thus tend to be relatively energy-inefficient. The group is focusing on small motors in residential areas, since these motors, which are associated with home heating and cooling units, run almost continuously.[93] From a perspective of global energy conservation, it becomes apparent that relatively small improvements in the efficiency of electric motors could have a significant, and continuing positive impact on global energy efficiency—thus helping reduce the impacts of human activities on the Global Biosphere.

Squeezing More Electricity Out of the Solar Spectrum

With the help of a new material, solar cells will soon be able to soak up energy from nearly the full range of the solar spectrum. This improvement is significant in that it could boost the efficiency of solar cells from a present best of 30 percent, to a factor of 50 percent, or more!

Solar cells utilize layers of semiconductors to absorb photons from sunlight, and convert them into electrical current. The problem is that each different semiconductor is capable of transforming photons only within a specific "bandwidth." The most efficient photovoltaic solar collectors consist of two different semiconductor layers stacked together to absorb light at different energies. Such collectors, however, are still only able to convert about 30 percent of the sun's energy into electricity. Although theoretical physicists had already calculated the two ideal bandwidths that would yield a maximum efficiency of 50 percent, until now, suitable semiconductor materials were simply not available.

A team of scientists at the Lawrence Berkeley National Laboratory in California appear to have discovered a material that is ideal for increasing the efficiency of photovoltaic collectors—a semiconductor called indium gallium nitride (InGaN). By varying the ratio of indium to gallium in the different layers of the photovoltaic collectors, they were able to "tune" InGaN's bandgap to match the criteria established by the theoretical physicists. Since the range of the bandgap can be designed to fit the solar spectrum exactly, a photovoltaic collector cell could use multiple semiconductor layers which are tuned to bandgaps across nearly the full range of the solar spectrum.

At a meeting of the Materials Research Society in Boston, scientist Wladek Walukiewicz and his research associates stated that studies on the new semiconductor material suggest it should be ideal for solar cells. Another major advantage of the new semiconductor compound is that it seems immune to effects from fault lines that tend to be created when different semiconductors are grown in adjacent layers. Such imperfections tend to disable some semiconductors, but blue lasers made from InGaN have been observed to function well, even when they are riddled with cracks and other defects. This resilient quality offers many exciting possibilities for prolonging the life of solar cells used in space, where wide temperature fluctuations and cosmic rays tend to damage the solar panels that supply power to the satellites. The scientific community had previously overlooked InGaN, because its bandgap was thought to be much smaller than Walukiewicz's results. It is now suspected that this discrepancy is due to the purity of the semiconductor, as the samples used by the Lawrence Berkeley group were made with a refined and prohibitively expensive methodology, to grow extremely pure crystals of InGaN—one atomic layer at a time. In the future, the research team intends to work with the National Energy Laboratory in Colorado, in a collaborative effort to develop a commercially viable and cost-effective process for manufacturing InGaN Photovoltaic Panels.[94]

Nuclear Photodeactivation Technology

In his book, *Earth Odyssey: Around the World in Search of Our Environmental Future,* author Mark Hertsgaard states: "Nuclear Fission represented the greatest power humans had ever tapped, but the associated costs and challenges were no less monumental. The embrace of atomic energy not only threatened the end of human civilization, it condemned humanity to environmental and health injuries that would take decades if not centuries to heal, and it saddled us with waste disposal responsibilities that for all intents and purposes will last forever."[95]

Despite this grim assessment, a unique new technology called "Nuclear Waste Photodeactivation," has been developed by Dr. Paul Brown, former President of Washington, DC-based Nuclear Solutions, Inc. This new process holds the key to "cleaning up" nuclear power generation technology. It also represents a practical means for eliminating the massive global stockpiles of spent nuclear fuel, dismantled nuclear weapons, and other forms of radioactive waste material.

In the Photodeactivation process, a linear accelerator bombards nuclear materials with high-energy x-rays, which effectively speeds up their normal rates of radioactive decay. Following this treatment, the radioactive materials which are targeted become stable, inert, and harmless, *within a period of days, or weeks,* rather than *thousands of years!* The relatively short time-span required for nuclear decontamination, thus facilitates the on-site storage of radioactive materials—until the treated isotopes can be rendered completely inert. As an added bonus, excess heat given off by the process can be used to generate electricity, which can be fed into any existing electrical power grid.

The Photodeactivation treatment plant is designed in modular components, which can be transported in three barges or flat-bed trailers directly to the waste site. This arrangement serves to completely eliminate the negative environmental issues associated with the transportation of nuclear wastes. If so desired, permanent installations for processing nuclear waste material can be added on, as an integral component of existing nuclear power plants.

This remarkable new technology thus offers a unique and environmentally appropriate solution for nuclear fuel decontamination, as well as an end to the expensive and environmentally damaging long-term storage of hazardous nuclear wastes. The process creates no toxic or hazardous waste products, and can treat nearly any kind of nuclear material (including weapons-related and medical wastes). The new technology has already been validated on an experimental scale. A pilot facility is presently being put into operation, and the process should be expanded to commercial scale within the next two years or less.

This innovative new technology is based on previously established scientific processes—applied in a new and unique way. From a global ecology perspective, the most fascinating aspect of the Photodeactivation process is that the basic concept is also theoretically applicable to the development of an entirely new generation of accelerator-driven reactor systems. Such new reactors have the potential for transforming the entire nuclear power industry into a safe, controllable, and environmentally appropriate technology.[96]

Table 1—Nuclear Photodeactivation Decay Intervals

Isotope	Normal Time to Become Inert	Time After Treatment
Cesium 137	302 Years	130 Days
Strontium 90	291 Years	Immediate
Iodine 129	1,700,000,000 Years	Immediate
Technetium 99	2,120,000 Years	43 Days

Fuel from Water

Two Canadian companies, Tathacus Resources Ltd. and Xogen Power, Inc., have combined resources to develop a new technology for producing commercial quantities of clean-burning hydrogen/oxygen gas fuel from ordinary water. The process offers the promise for creating a future scenario, where atmospheric pollution from fossil fuels could be reduced to a fraction of what it is today!

The prototype hydrogen-fuel generator is said to consume relatively little electricity, has no moving parts, and uses no chemical additives. The process is cost-effective, in that commercial volumes of hydrogen can be produced *in-situ*—thus effectively eliminating the costs and technical problems normally associated with transportation and storage of conventional fossil fuels.

The companies are working cooperatively to adapt the technology for both home and commercial applications, and were expected to complete their first self-contained home furnace by late 2001. The Xogen Beta Prototype Generator separates the oxygen and hydrogen atoms in water, delivering them in a suspended gaseous state. Power is drawn from a 24-volt battery, which can be charged using solar panels. According to reports from company officials, electricity generated from the burning gas somehow functions within an "over-unity modality" in that *the gas produces far more energy than is initially used to create it.*[97]

Brown's Gas: Another Revolutionary Fuel-From-Water Technology

Brown's Gas is essentially water, split into its two components—hydrogen and oxygen. Through a process called alkaline electrolysis, the two gases are combined under pressure in a proportion of 2:1, which facilitates effective combustion. When ignited, the gas recombines by implosion back into water—collapsing into a vacuum-to-water ratio of 1,886.6 to 1. Due to the extensive research already conducted with this technology, the various aspects of the process have been subjected to detailed analysis, and have yielded a variety of applications which have proved to be practical, and ecologically appropriate.

Brown's Gas was discovered in the 1970's by Australian Yull Brown. Brown's Gas Generators were first developed and manufactured in Australia, but the technology was subsequently transferred to China, where the generators were mass-produced. In addition to its initial applications for welding and brazing, the original concept was eventually expanded to include a variety of other applications such as water desalination, destruction of medical and toxic wastes, medical applications, and materials hardening. In 1996 the Chinese invited Brown to build a Brown's Gas System for use in automobiles. Due to poor health, however, he returned to Australia, where he eventually died.

Through a collaborative association between Yull Brown, and the Planetary Association for Clean Energy (PACE) in Ottawa, Canada, arrangements were made to produce generators and other applications that would comply with North American and European standards. One novel application which has emerged from this partnership is a new technology for processing heavy crude oil and oil sands. Other promising uses for Brown's Gas technology include applications as a gasoline substitute for automobile engines, and as a means to optimize combustion of fossil fuels such as wood, coal, natural gas, and oil—with the objective of achieving complete combustion with minimal environmental pollution. Research is also underway to develop new applications for storing energy from hydroelectric, wind, and solar power; producing Brown's Gas by electrolysis during periods of low demand, and burning it to produce electricity during periods of high demand. So far, the possibilities look very promising.

Considering the abundant supplies of water on our planet, Brown's Gas offers excellent possibilities as the best "Carrier" for alternative energies which we have at this time. Since it emits only pure water as a by-product of combustion, Brown's Gas is non-polluting. Of even greater importance, is the fact that the combustion of Brown's Gas *does not consume Atmospheric Oxygen,* as is true with **both** hydrogen and fossil fuels. Neither does it emit nitrogen oxides (as with fossil fuels and hydrogen). Like hydrogen gas, Brown's Gas is adaptable to a variety of existing energy technologies—without requiring any major modifications. Other applications include heating for cooking and living spaces, and cooling for homes and workspaces. A Brown's Gas system might also provide the ideal complement for solar photovoltaic systems, by replacing the need for cumbersome banks of heavy storage batteries, which are costly, require regular maintenance, and represent the environmentally inappropriate aspects of solar technology.

Brown's Gas generators use a mixture of sodium hydroxide and water to form an electrolyte, which has a conversion efficiency of 90-95 percent (excluding cable and system losses). Whereas theoretical energy production for regular hydrogen-oxygen gas is about 50,000 BTU's per pound, Brown's Gas produces approximately 66,000 BTU's per pound. According to PACE Founder and President, Dr. Andrew Michrowski, if combined with existing proprietary technologies, this output could be theoretically increased to as much as 210,000 BTU's per pound! If even 80 percent of this energy were captured, this unique technology could alleviate the two major problems with alternative energy systems such as solar, wind, and tidal power—variable input and energy storage. Brown's Gas can also be utilized to increase the efficiency of hydrogen fuel cells by providing a cheap and easily transportable source of hydrogen.

Yull Brown conducted experiments with a variety of automobiles, with different types of internal combustion engines; all carefully monitored with the appropriate scientific instrumentation. At one point in his research, while being officially monitored, *he drove 1,000 miles using only a single gallon of water as fuel!*

A major advantage of this fascinating technology is that conventional internal combustion engines require little modification to run on Brown's Gas. The main modification is simply replacing the carburetor with a pressure reducer and throttle valve. It is also necessary to re-tune the engine to compensate for the high flame speed of the hydrogen/oxygen fuel mixture. A second major improvement associated with converting from gasoline to Brown's Gas involves the extension of engine life. Since the only product of hydrogen-oxygen internal combustion is water, there is no carbon deposited on either the spark plugs or valves, and no corrosion in the exhaust lines from acidic vapors (as with conventional petroleum fuels).

The cells of a Brown's Gas generator produce approximately 340 liters (13.6 cu. ft.) per kilowatt hour, which is estimated to be between 7 and 58 times less expensive than with bottled acetylene (depending on electric rates and bottling costs). For purposes of comparison, operational cost for a small gasoline car in Ontario is approximately 2 cents per kilometer, and for an electric car about 1 cent per kilometer. With Brown's Gas a full-size car can be operated for approximately .13 cents per kilometer (in Canadian currency)! At this stage of development, for automotive use, Brown's Gas can be stored in conventional gas bottles. Despite the fact that the energy-to-weight ratio for Brown's Gas is only about one-third of that for gasoline, this still represents a significant improvement over batteries.

Brown conducted a series of experiments in which he determined that large amounts of the gas could be stored conveniently in relative small volumes. He even envisioned that bottles of Brown's Gas could be used to run cars, trucks, and even home electrical systems. Ongoing research in Canada indicates that it is feasible to install small battery-charged Brown's Gas units (about the size of a 2-liter soda bottle) which would essentially function as "mini onboard hydrogen generators" for motor vehicles.

In the opinion of Dr. Michrowski, Brown's Gas is one of the most promising new solutions available as an environmentally appropriate "Fuel for the Future." In spite of the fact that the technology is still largely unknown to most North Americans, it is practical, has already been tested under commercial conditions, and has a number of inherent safety features. Brown's Gas is relatively inexpensive, both as a primary fuel, and in terms of capital requirements. The technology is also ideal for converting conventional industrial power plants, so that clean electrical power could be cheaply and efficiently produced, with minimal impact on the environment.[98]

A Hydrogen Future

In his novel, *The Mysterious Island*, Jules Verne created a world where he envisioned that water would be "the Coal of the Future." Today, the idea of using hydrogen (the most abundant element in the universe) for energy is now finally being taken seriously.

Several islands, where oil prices are high, have already established themselves as prototypes for a hydrogen energy transition: Hawaii, Iceland, and Vanuatu. The major energy corporations have created hydrogen divisions, and the big auto manufacturers plan to market the first commercial fuel cell vehicles between 2003 and 2005. Japan alone, plans to spend four billion dollars on its World Energy Network Program by the year 2020!

A transition to hydrogen power still faces a number of technical, economic, and political obstacles. The major problem at this point in the evolution of the technology, is the lack of a viable hydrogen infrastructure.[99]

> **Necessity is still the mother of invention,**
> **and we're on the brink of a veritable explosion**
> **in technical engineering inventiveness.**
> **After all, Energy is wonderful stuff,**
> **liberating and wealth-creating.**
> **We want more of it,**
> **but without the damaging side-effects**
> **that now threaten to undermine**
> **the prospects of the whole of Humankind.**
>
> ~ Jonathon Porritt, *2000—Playing Safe:*
> *Science and the Environment*, p. 74.

The Next Step: Combining Oil and Hydrogen
Technologies

Research by Nobel Prize Nominee, Dr. Ruggero Maria Santilli, would indicate that the use of hydrogen fuel **will not** resolve our present environmental problems, due to excessive emissions of carcinogens and carbon dioxide. Santilli's research also suggests that burning hydrogen, which is produced by regeneration or the electrolytic splitting of water, using electricity from fossil fuel power plants would most likely create an alarming state of atmospheric oxygen depletion in the Global Biosphere. He points out that the burning of gasoline releases higher percentages of carcinogens and other toxic substances than with any other fuel. He goes on to say that when fossil fuel combustion occurs on a large scale, green vegetation is basically overwhelmed, and thus unable to entirely recycle the disproportionately large amounts of atmospheric CO_2 a, factor which has been mainly responsible for creating the present "Greenhouse Effect." According to Dr. Santilli, in 1950 the atmospheric percentage of CO_2 at sea level was approximately .033 percent. In 2000, however, measurements taken at his South Florida laboratory indicated a *thirty fold increase* over the 1950 figure!

In Dr. Santilli's opinion, however, it is a lesser known problem associated with fossil fuel combustion that represents the most serious threat to the Global Biosphere. He contends that when hydrogen is burned, atmospheric oxygen is transformed into water—which effectively removes this life-essential gas from the atmosphere. He feels that hydrogen combustion on a reasonable scale does not necessarily constitute a serious environmental problem. On a massive global scale, however, hydrogen combustion (especially if the gas is produced by regenerative methods) could prove to have potentially catastrophic effects on the environment simply because Oxygen is "the very foundation of life." He adds that in the extreme event that hydrogen of regenerated origin was used to completely replace fossil fuels, Earth would be rendered completely uninhabitable within a short period of time!

In Dr. Santilli's opinion, *the very existence of the Greenhouse Effect provides irrefutable proof for the case of atmospheric oxygen deletion* since what we are dealing with now relates directly to that amount of CO_2 in the atmosphere which has *not* been converted back to oxygen by photosynthetic plants. Santilli points out that back in 1950, the basic atmospheric concentration of oxygen at sea level was slightly under 21 percent. During the year 2000 he claims he routinely measured local oxygen depletions which were three-to five-percent lower than this baseline value. He goes on to suggest that oxygen concentrations could prove to be even lower in densely populated cities like New York, London, or Tokyo, with the problem being magnified as elevation increases.

From an environmental perspective, Santilli states, "... whether used for direct combustion or in fuel cells, hydrogen produced with regeneration methods (e.g. from natural gas) does avoid the release of carcinogenic substances and carbon dioxide in the exhaust, but causes an alarming oxygen depletion, which is considerably greater than that caused by fossil fuel combustion under the same energy output. This depletion is due to the fact that gasoline combustion turns atmospheric oxygen into CO_2, part of which is recycled by plants into O_2, while hydrogen combustion turns atmospheric oxygen into H_2O. This latter process permanently removes oxygen from our atmosphere in a directly usable form, due to the excessive cost of water separation to restore the original oxygen balance."

Thus, if used only in modest quantities, the burning of hydrogen would not cause significant environmental harm. However, on national and global scales, burning hydrogen produced through regenerative methods would most likely have a catastrophic impact on the global environment, since oxygen is such a basic necessity for most life on earth. Santilli reiterates, that should hydrogen produced using regenerative processes be used to completely replace fossil fuels, our planet would be rendered uninhabitable within a relatively short time—due to the *permanent removal* from the atmosphere of 76 percent of the oxygen that is currently consumed in the burning of fossil fuels.

Burning hydrogen, produced from the electrolytic separation of water, creates its own set of environmental problems. Thus, the most environmentally sensible way to use hydrogen is to derive it from water separation technologies which use electricity derived from benign alternative energies (such as wind, solar, hydroelectric, wave, or tidal power). The overriding implications are that most of the major environmental problems we presently face directly relate back to the antiquated fossil fuel technologies used for generating electrical power. These outdated applications need to be revised and cleaned up, with new applications of sustainable power technologies.

In order to resolve these complex interrelated problems, and to create a cheap environmentally acceptable method of electrical power generation, Dr. Santilli has developed a new technology which upgrades hydrogen gas into a new combustible gas, which he has named "MagneGas." MagneGas is produced using a new technology called PlasmaArcFlow, which involves flowing liquids through a submerged electrical arc. This process transforms the molecules of the liquid into plasma at 7,000°F, which is composed mainly of ionized H, O, and C atoms—plus solid precipitates. These ionized atoms recombine into a combustible gas of a new (and as yet unknown) chemical composition.

MagneGas has several unique qualities, among which is the fact that its chemical structure can not be properly analyzed with conventional mass spectrometric gas chromatography, since it is apparently composed of large clusters, ranging up to 1000 a.m.u. in molecular weight. The gas is equally unanalyzable with infrared detectors, since the clusters in the MagneGas exhibit no IR signature at all. In addition, the IR signatures of normal CO and CO_2 molecules apparently also become "mutated," since they register atypical IR peaks, which would suggest the formation of new internal bonds. The combination of these aberrant features serves to validate the fact that MagneGas has an energy content which is considerably higher than would be predicted by quantum chemistry. It apparently stores energy at three different levels: clusters (i.e. magnecules), molecules, and the newly formed internal molecular bonds—a phenomenon which Dr. Santilli compares to a multistage rocket.

Since MagneGas defies analysis using conventional scientific methodology, Dr. Santilli and his professional colleague, D. D. Shillady, have created an entirely new theoretical scientific framework which they call "Hadronic Chemistry." Analytical scans of the same MagneGas sample of over time indicate the presence of different types of magnecules—a phenomenon they call "Magnecule Mutation," where the observed discrepancies are apparently due to collisions between magnecules, resulting in fragmentation and recombination with other magnecule fragments.

Research has also shown that the percentage of hydrogen depends on the liquid which is used for its production; with the highest percentages of hydrogen being produced from crude oil! A key feature of the new chemical species of magnecules formed during the PlasmaArcFlow process is that they embody important implications for the future of fuels. When subjected to a sufficient degree of magnetic polarization, hydrogen can apparently acquire sufficient energy density to eliminate the need for liquefaction. This fact alone would imply significant advantages in terms of cost, production, storage, delivery, and other basic approaches to using hydrogen. When subjected to sufficient magnetic polarization, hydrogen can be made to yield an energy output equivalent to that of gasoline, as has already been demonstrated by researchers at U. S. MagneGas, Inc., using a conventional automobile which can run on either gasoline or compressed MagneGas. Thus the new chemical species (i.e. Magnecules) present in MagneGas could effectively serve to eliminate the basic power loss in the shift-over from gasoline to alternate fuels.

One of the most exciting aspect of this new technology is the fact that the PlasmaArcFlow reactors used for producing MagneGas have been independently calibrated to function at a commercial over-unity of at least six to three, using liquid waste instead of water. In other words, for each unit of electrical energy used, the reactors produce up to six units of energy—a combination of the energy embodied in the MagneGas plus heat. The additional five units of energy originate from the liquid waste. In short, MagneGas reactors are apparently capable of tapping energy *at the molecular level,* in a process which is analogous to the way nuclear reactors tap energy at the nuclear level. Dr. Santilli emphasizes the advantages of the large commercial over-unity typical of MagneGas reactors, when compared with the under-unity of conventional hydrogen production technology—which rarely exceeds a value of 0.8. Thus MagneGas represents an effective way to significantly reduce the costs of hydrogen production, with the added advantage of bypassing the liquefaction stage, while still yielding a power output similar to that of gasoline.

From an environmental standpoint, the unusual chemical composition of MagneGas offers the major advantage of being able to synthesize oxygen-rich fuel from liquid wastes, rather than directly from the atmosphere. Perhaps the most significant advantage lies in the fact that the burning of MagneGas actually results in a *positive oxygen balance,* whereby *more* oxygen is produced than is used during the combustion process! By transforming hydrogen into MagneGas, the following results could be accomplished: First, the oxygen depletion typical of hydrogen combustion would be transformed into a positive oxygen balance. Second, significant reductions in CO_2 emissions would be achieved with MagneGas (when compared with fossil fuels), while still maintaining a positive oxygen balance (Dr. Santilli also claims that MagneGas Exhaust meets the most stringent government regulations even *without* a catalytic converter).

In an experiment comparing the exhaust emissions of Magne-Gas with Gasoline, a 1999 Honda Civic was run using both fuels. With conventional gasoline, exhaust emissions were compared with those of the same vehicle using MagneGas. Results were as follows:

Table 2: Emissions of Gasoline vs. MagneGas

Hydrocarbons—0.234 g/mi. = 900% × emissions of MagneGas.
Carbon Monoxide—1.965g/mi. = 750% × emissions of MagneGas.
Nitrogen Oxides—0.247 g/mi. = 86% × emissions of MagneGas.
Carbon Dioxide—458.655 g/mi. =195% × emissions of MagneGas.

These results clearly demonstrate the environmental Superiority of MagneGas, in comparison to conventional gasoline as a fuel for automotive transportation.

Additional factors which would favor the use of MagneGas over pure hydrogen as a fuel are as follows: 1) MagneGas is cost competitive with fossil fuels, because it can also be produced from liquid wastes, as an income-producing by-product of recycling, 2) MagneGas, with its energy content of about 800-900 BTU/cu. ft., significantly exceeds that of hydrogen (about 300 BTU/cu. ft.), due to the newly discovered energy bonds embodied in the magnecules, 3) Since the desk-sized reactors are compact and portable, MagneGas can be produced in virtually any location, at rates approximating 1,500 cc/hr—an amount sufficient to run a compact car for about three hours of city travel, 4) Due to the attraction between magnecules, MagneGas can easily penetrate semi-permeable membranes, 5) MagneGas is suitable for all conventional fuel applications, which include cooking, metal cutting, and automotive applications, 6) MagneGas can also be used in fuel cells, thus extending the environmental advantages of this new power generating technology.

Since crude oil is ideally suited for processing with PlasmaArcFlow reactors, MagneGas Technology would appear to represent one of the best alternatives for hybridizing crude oil and hydrogen technologies. Using this approach, crude oil can be transformed into a fuel which burns much cleaner than gasoline, at a significantly lower cost, with a much simpler and more efficient technology than is used in conventional oil refineries. Moreover, the fuel thus produced is composed of over 50 percent hydrogen. In Dr. Santilli's own words, "...Crude Oil, Hydrogen, and fuel cells remain indeed fully admissible in this new era of environmental concern, provided they are treated via a basically new technology whose quantitative study requires a new chemistry, Hadronic Chemistry."[100]

Aneutronic Energy: A Non-Radioactive, Non-Proliferating Nuclear Fusion Power Technology

According to a report by a Special Committee of the U.S. National Research Council, the world may be only a step away from developing an effective new nuclear power technology. This technology is non-radioactive, and uses a fuel that eliminates any possibility of converting the fuel or its by-products into weapons of mass-destruction. These conclusions were reached in 1987—over a decade ago—by a group of scientists at the First International Symposium on the Feasibility of Aneutronic Power, held in 1987 at Princeton University's Institute for Advanced Study.

Aneutronic Energy Technology uses nuclear reactions that involve non-radioactive nuclei—both as reactants, and as reaction products. Nuclear reactions are classified as "Aneutronic" if less than one percent of the total energy released is carried by neutrons, and if less than one percent of the reactants (fuel) and reaction products (waste products) are radionuclides. The end product of aneutronic reactions in all cases is mainly helium, a non-radioactive inert gas.

In 1982, an Experiment referred to as "Migma IV" was carried out, using a fuel density *1000 times less* than for the best Tomak Fusion Reactor. The temperatures that were produced exceeded Tomak's best *by a hundredfold.* Since Reaction Confinement Time was *15 times longer* than the best results with Tomak Reactors, this means that the product produced by the Migma IV was approximately *1500 times greater in magnitude,* than from the best Tomak results! Due to a combination of academic dysfunctionality, and the cumbersome funding politics for the extravagant and ultimately fruitless. Tomak research project, the Migma Program (a promising, cost-effective, benign nuclear power technology) was overshadowed, and finally pushed aside. Over the past 30 years, the Western World has lavished some *ten billion dollars* on conventional plasma fusion research. By contrast, only *23 million dollars* was spent on the Migma Program over a ten-year period.

Commenting on this glaring disparity and the massive waste of public dollars in funding conventional plasma fusion power vs. aneutronic nuclear power, the U.S. Senate Appropriations Committee declared in 1982, "To date, basic research in the field of nuclear fission and fusion has largely overlooked the potential for aneutronic nuclear alternatives using light metals such as lithium, that produce no radioactive side effects. The committee recommends that the DOE give higher priority to this non-radioactive and non-proliferative nuclear potential." [Apparently this poignant advice was simply ignored.]

Since aneutronic reactors do not require heavy shielding, as with conventional nuclear reactors, they have a relatively high power-to-weight ratio, and a low fuel weight (*about 100,000 times the concentrated fuel energy of non-nuclear fuels*). Costs for aneutronic reactors are about ten percent lower than for "dirty fusion" uranium reactors. Other positive aspects include reasonable fuel availability; significantly lower capital costs for power plants; no waste heat pollution; and the economic advantages of being able to create modular units as small as one megawatt.

Other feasible applications for aneutronic reactors include their use as power supplies for remote radar and telecommunications installations. The smallest aneutronic power plant (30 Kw), about the size of the proposed Migma V unit, would be suitable for powering such installations. Aneutronic power units should also prove ideal as non-polluting marine propulsion applications for passenger, freight, and military vessels. Small aneutronic reactors can apparently even be created to produce a few megawatts of electricity (in contrast to the minimum practical size for fission plants—or the projected fusion power plants). Such small power units open up possibilities for mass production, which should result in far lower capital costs per kilowatt, than for conventional nuclear reactors, which are built one at a time by large contracting companies. Mass production of aneutronic reactors would also provide distinct advantages for developing countries, and for small communities in developed countries, where initial capital costs represent a major barrier for building conventional power plants. Add to this the major environmental advantages of using non-radioactive fuel, producing non-radioactive waste, and of not having to deal with environmental impacts related to the waste heat typically produced by conventional nuclear reactors.

One of the most compelling reasons for continuing to develop aneutronic energy is that aneutronic reactors *cannot* breed plutonium for use in nuclear weapons. Since radioactive fuel, radioactive waste, and heat pollution are the major political and environmental reasons *against* developing nuclear power, aneutronic energy offers us exciting possibilities for developing a clean, and environmentally appropriate energy technology—a technology which would benefit a wider range of global society, while reducing the impacts of human technology on the Global Biosphere.[101]

"Zero-Point Energy" Device: A Renewable, Pollution-Free Power Source for the Future

On March 26, 2002 United States Patent #6,362,718 was granted for "The Motionless Electromagnetic Generator" (MEG). This revolutionary new device is scheduled to become the first commercially available "Free Energy Device" in history. The new device provides free electricity from the vacuum. Since it has no moving parts, it is expected that the device will be capable of providing reliable energy for at least several years—without interruption.

The realization of the new MEG device is particularly significant, since it effectively validates the new Science of Scalar Magnetics. Historically, the U. S. patent office has regarded such over-unity devices, which appear to "get something for nothing," with great skepticism. A new vanguard group from the mainstream scientific establishment has apparently finally rationalized its thinking to accommodate the MEG device, which within the laws of scalar magnetics *does not* break the law of conservation of energy, but recognizes the fact that the energy *is conserved* within the fourth dimension (time), which apparently exists outside of our ordinary three-dimensional universe.

The new MEG device consists of heavy magnetic coils, and an electronic controller unit. It "generates" electrical energy by tapping longitudinal (scalar) electromagnetic waves, which apparently exist in virtually infinite abundance in the vacuum of space. This "Infinite Ocean of Energy," which permeates all matter and space is often referred to as "Zero Point Energy," since it prevails even at temperatures of absolute zero.

One of the four inventors credited with bringing the new MEG device into practical reality is Tom Bearden, who details the operation of his device in a scientific paper on his website entitled, "The Motionless Electromagnetic Generator: Extracting Energy from a Permanent Magnet with Energy Replenishment from the Active Vacuum." (www.cheniere.org). A new company called Magnetic Energy Limited has been set up to produce the devices, and a manufacturing plant is currently being constructed in an "unnamed friendly nation." The MEG units are designed to produce free electricity indefinitely—with little or no maintenance. Individual units, designed to produce 2.5 kilowatts, can easily be connected together to provide additional wattage. After an initial production period, larger models, designed to produce 10 kilowatts, will be produced. With two of these 10kw units, a typical household could thus become entirely independent of the large centralized commercial energy grids.

Inventor Bearden comments, "I will admit that the chief scientist of an experimental group in a large company was rather stunned at the type of output we were able to obtain. The MEG may look just like a transformer, but it is not. It is a completely different breed of cat." Bearden also emphasizes that the new MEG device utilizes newly discovered "Longitudinal Electromagnetic Waves" of the vacuum.

Since patents are not granted for inventions that do not work, in a sense, the awarding of a U. S. patent for the new device represents the dawning of an entirely new era of clean, environmentally sustainable energy for the human race. The very existence and function of the MEG device validates the fact that energy is abundantly available anywhere—for free. We only need to build the appropriate devices to tap into this resource. MEG devices will make energy available in very remote locations, and could be installed in electric cars to produce a truly "Fuel-Less Automobile!" By reducing, and eventually eliminating, our dependence on fossil fuels, with their traditional terrorist-vulnerable centralized power plants, and associated power distribution infrastructures, we would end up with inexpensive, pollution-free, efficient, decentralized power systems, which would be ecologically compatible with the natural environment.

Perhaps the most astonishing aspect associated with the discovery of the MEG device, is the discovery that time itself is really compressed energy, and that the free energy is actually coming from the so-called "Time Domain"—that sea of longitudinal electromagnetic waves that fill the vacuum of space-time. It thus now appears that time itself is energy, compressed by the same factor that matter is compressed energy—the speed of light squared. Thus, Einstein's famous equation, $E = mc^2$ has a new companion equation, $E = ATc^2$, where delta-T stands for change over time.

The implications for the commercialization of free energy are nothing less than earthshaking! Bill Morgan, author of an article entitled, "Bearden for Beginners," states, "If we threw as much money at this technology as we are spending on the oil wars we would be free of the need for oil in less than a decade. With fuel-less cars, air pollution would be greatly lessened. Third-world nations can raise their standards of living eventually. And the energy is free. It never runs out."[102]

Basic Strategies for Clean, Renewable Global Energy Production:

1) A Major Global Shift is needed away from fossil fuels like wood, coal and oil, to cleaner-burning fuels—and eventually to

completely non-polluting renewable energy sources. Cleaner burning fuels would include: natural gas, hydrogen, Brown's Gas, and MagneGas. Meanwhile, non-polluting, renewable energy resources such as solar, wind, wave, tidal, and geothermal power should be implemented on a worldwide basis. In the transition to alternative energies, it is important to factor in any pollution created by those power sources used to charge electrical vehicles, or to generate hydrogen via electrolytic processes. A major overriding consideration in the burning of any fuel is the possibility of global atmospheric oxygen depletion; resulting from the burning of fossil fuels beyond the point where the natural photosynthetic systems of planet Earth can process the resulting carbon dioxide to regenerate sufficient oxygen to sustain the balance of nature.

2) New, clean-energy technologies need to be brought into commercial reality as soon as possible. In many cases, existing utilities can be cleaned up or converted, to provide inexpensive, dependable, clean energy for their consumers. Concurrently, major efforts should be mounted to make the shift from centralized power generation, to compact, efficient, and clean energy generating systems, which could be installed directly in homes, institutions, and industries. This arrangement would eventually serve to eliminate power distribution grids altogether, with their associated costs, maintenance, vulnerability, environmental incursions, and transmission inefficiency. Different types of alternative Energy technologies can also be combined to create hybrid systems which would provide clean and reliable power.

3) A new economic model for electrical power generation needs to be established, to obtain the necessary investment for technically feasible mega-projects like the Solar Tower Generating Plant in Australia, or the Hot-Rock Geothermal Project in Cornwall, England. Many of these new alternative power-generation technologies tend to have *large* capital costs, but *low* maintenance costs, and *no* fuel costs. (This represents a major conceptual shift from fuel-intensive, to capital-intensive systems).

4) The development and distribution of new energy conservation technologies needs to be stimulated through financial incentives. In this scenario, energy-efficient devices would be rewarded with economic incentives, and energy-inefficient devices would be correspondingly penalized.

5) The development of entirely new power-generation technologies should be supported through major funding programs, sponsored by both government and private sectors. These programs should include new applications of "Zero-Point" energy devices, and any other non-conventional alternative energy technologies which function outside of established physical parameters (A prime example of this type of technology is the new MEG device, which required that a new system of "Scalar Physics" be developed in order to validate the patenting process, and explain how the device is able to function).

6) Wasteful energy projects such as conventional plasma fusion (e.g. the Tokamak Project, etc.) have frittered away billions of dollars of public tax money over the past 30 years—with little, if any, positive result. Such "Dead Elephant Projects" should be abandoned, and their personnel transferred into programs for developing new technologies, which will provide exciting challenges for their technical expertise. The U.S. Department of Energy and corresponding agencies in the major nations of the world need to work together to support programs such as aneutronic energy, which offers the promise of controlled nuclear fission and fusion reactions with *no* radioactive by-products—thus completely eliminating the possibility for converting spent nuclear fuel into weapons of mass destruction (as *is possible* with most conventional nuclear reactors).

7) In order to successfully integrate these new clean alternative energy technologies into the fabric of global society, without incurring major economic upheavals, the big multinational energy companies should be provided with fair opportunities to compete for the manufacturing and distribution rights of the new energy generating units for customers throughout the world. By making the transition to clean, inexpensive, and portable energy

generating units (similar to furnaces or air conditioners), developing nations could afford to bring cheap and abundant electrical power to even the most remote areas of the world. Shifting away from fossil fuel dependence and eliminating centralized national electrical grid systems would drastically reduce our dependency on oil, and eliminate a major source of global conflict, also reducing our vulnerability to terrorist attacks. Our remaining global oil resources could then be put to far better use as valuable feedstock resources for producing key components for the industrial technologies of the future.

METASTRATEGY V

A GLOBAL "GREEN REVOLUTION"

**If we people of Earth
are to avoid a massive disaster
within the lifetime of our children,
our most critical and urgent task
is to bring forth
a transformed vision of progress,
one of sustainable and replicable development.**

~ Gerald O. Barney, *Global 2000 Revisited,* p. 5.

A Global Climate Threshold?

In 2001 the United Nations Governmental Panel on Climate Change (IPCC) predicted that Earth's average surface temperature would rise as much as 10° F during the 21st Century. Recent scientific data now suggest that this forecast may have been even too conservative.

During the winter of 2002 global temperatures increased by an average of 4.3 percent—the largest jump in recorded history! The National Academy of Sciences now cautions that the concept of a gradual temperature rise may prove to be false. Instead, a dramatic and abrupt climate shift could possibly be triggered *within the space of only a few years.* This type of abrupt shift could result in large-scale extinctions of plant and animal life on Earth. For example, at the end of the so-called "Younger-Deyas Interval" some 11,500 years ago, the global climate shifted abruptly, and within the space of only a few years.

With carbon dioxide emissions expected to double over the next century, there is evidence to suggest that the Global Biosphere may be shifting toward a "Threshold Event" that could plunge global weather systems into chaos in less than a decade. Within

this context, Author Jeremy Rifkin writes, "If the U.S. continues
to refuse to act on global warming, we may have already have
written our own epitaph."[103]

A Vision for the Commons

I walk into the Commons,
not really having a destination.
The Commons is where
my basic community needs are satisfied;
to see and be seen, to connect with others,
to know that we are part of the same community.

~ Sharif Abdullah, 1999, *Creating a World that Works for All*, p. 97.

Earth's Natural Resources on "Overdraft"

A recent study by the environmental organization, Redefining
Progress, has determined that the consumption of Earth's natural
resources—mainly forests, land, and energy—presently exceeds the
rate at which they can be replenished and could thus send the planet
into a state of "Ecological Bankruptcy." Program Director for the
non-profit organization, Mathis Wackernagel, explains that Earth's
resources "are like a pile of money anyone can grab while they all
close their eyes—but then it's gone."

According to environmental scientists, in the past few decades the demand for natural resources has risen to a level at which
1.2 years will be required for the Global Biosphere to regenerate the
resources that are removed in a single year. In 1961, for example,
the global national resource demand equaled only 70 percent of the
yearly replacement interval. Since that time this public demand
figure has continued increasing steadily. Wackernagel puts this
increasing "Eco-Debt" into perspective by stating, "If we don't live
within the budget of nature, sustainability becomes futile."

The redefining progress study details the impacts of human
activities (the "Ecological Footprint") on marine fisheries, timber harvesting, building infrastructure, and the burning of fossil

fuels relative to carbon dioxide emissions. The researchers combined figures from government statistics and other sources, to come up with the estimates for just how much land would be required to meet human demands for these actions. For example, the group determined that in 1999 each individual human consumed an average which was equivalent to 5.7 acres of land. The global averages proved to be significantly lower than for industrialized nations like the U. S. and the United Kingdom, which consumed 24 acres and 13.3 acres per person, respectively.

Due to lack of complete data, the study excluded several natural resources such as local freshwater use, and the release of solid, liquid, and gaseous atmospheric pollutants (other than carbon dioxide). Wackernagel feels that results from his research could also be useful for evaluating the impacts of new technologies on the environment. By using alternative technologies, which produce renewable energy from natural biological processes, society could enhance the general quality of life, while reducing the negative impacts on the environment. In addition, governments would be able to more accurately assess the impacts that consumers and businesses have on environmental resources, and could then work to develop new approaches for reducing consumption. Says Wackernagel, "Like any responsible business that keeps track of spending and income to protect financial assets, we need ecological accounts to protect our national assets. And if we don't, we will have to prepare for ecological bankruptcy."[104]

A "Global Green Deal"

Considering the major challenges presently facing the Earth's Ecosphere, what is needed is an entirely new and transformative global environmental program, a program designed to renovate virtually all aspects of human society—including rich and poor countries alike. With appropriate financial incentives, supported by leadership from government and industry, the development of sustainable "Green Technologies" could be dramatically accelerated— as was the case with computer and Internet technologies.

Although some progress has been made in selected areas, our major environmental problems like global warming, atmosphere and water pollution, loss of arable land, deforestation, water scarcity, and species extinction are all worsening, and at an accelerated pace; with poverty and overpopulation at the forefront of this process of environmental deterioration.

Xerox, Compac, and the 3M Corporation provide examples of the many firms that have already realized that "Green Technologies" can be "Profitable Technologies." By cutting their greenhouse gas emissions by half, these companies have increased their revenues correspondingly by as much as 50 percent! To be successful, any proposed global green deal must be sure to include "Ecologically Critical" countries such as China and India—with their relatively huge populations and ambitious plans for industrial and social development. Otherwise, they alone could destroy the balance of the global ecosystems, and thus doom the rest of the world to an irreversible trend of severe global warming.

Although we must change our personal and commercial lifestyles profoundly, most of the required technological elements already exist. It is only a matter of bringing them up to industrial scale and integrating them into the existing global infrastructure, in order to open bold new pathways which will lead to sustainable futures. Governments would need only to shift their existing subsidies from supporting antiquated environmentally incompatible technologies, into new sustainable "Green Technologies." If even 50 percent of the 500-900 billion dollars of global subsidies could be redirected into the development and application these sustainable green technologies, we would have taken a major step towards stabilizing the balance of the Global Biosphere.

Although the proposed "Greening Program" would not provide Permanent solutions to our present environmental problems, it would give us time to make positive changes, and thus allow global society an opportunity to make the transition to environmentally cleaner industrial systems and more ecologically appropriate lifestyles.

For this "Global Green Deal" to become a reality, a profound shift in global consciousness will be required. The resulting "Payback," however, would be well worth it, as the benefits would cut across class and national boundaries to improve the health of the Global Biosphere for everyone on Earth.[105]

Donating More than Money

When world banking multinational HSBC recently donated 50 million dollars to conservation programs, this was, in itself, truly a landmark achievement. What is more important, however, is that the company has also set a new trend, which conservationists expect will further magnify the bank's generous contribution to the environment. In addition to contributing $16 million to Earthwatch Institute for training scientists in developing countries, and funding conservation projects around the world, HSBC has created a Network of 2,000 of its own employees to serve as "Environmental Ambassadors."

The "Investing in Nature Program" is a $50 million, five-year partnership funded by HSBC to work with Botanic Gardens Conservation International, Earthwatch, and the World Wildlife Fund. The overall objectives of the project include the protection of 20,000 species of plants from extinction, restoration of some of the world's major river systems, and a plan to send scientists and HSBC fieldworkers to various strategic environmental sites throughout the world (http://www.hsbc.com/1/2/sustainability/environment).

According to Shawn Fitzgibbon, Director of Corporate Programs for Earthwatch, the fact that HSBC is lending its own personnel to implement the program, serves to magnify the effectiveness of the conservation program. It is hoped that program participants will achieve a much greater understanding of the issues at hand, which will, in turn, result in increasing the momentum for positive environmental change within their own corporate culture. Says Program Director Fitzgibbons, "It is a tremendous opportunity for the people who are participating. In return, HSBC is developing a staff that understands, and has experienced first-hand, some of the environmental issues that are affecting the world and will affect the bank. It is positioning them to do well in the long term financially, socially, and environmentally."

The 2,000 HSBC employee network intends to create its group of environmental ambassadors from within the bank. These selected individuals will be provided with grant funding for local conservation projects, when they return to their communities. Earthwatch has already established similar partnerships with several other companies including Starbucks, Mitsubishi International, and Royal Dutch Shell. Steve Volkers, Director of Corporate Programs for the nature conservancy, is a firm believer in getting corporations directly involved with conservation projects. Says Volker, "We look at long-term solutions to protect biodiversity, and we need partners to get this done. No single organization can do this alone. The mutual interests become pretty clear, and it is pretty easy to find where the win/wins are and where they aren't."

The practice of establishing cooperative partnerships between corporations and conservation organizations has become increasingly common in recent years. For example, the nature conservancy was recently funded by a $100,000. Grant from pharmaceutical giant, Bristol-Myers Squibb, to develop ways and means for protecting the ecosystems of Puerto Rico. In this partnership, company volunteers will help with workshops, communications, and public relations. Ideally, such corporate-environmental partnerships should generate additional interest and support for the continued expansion of existing environmental programs.

In a similar alliance Mead Corporation and its subsidiary, Escanaba Paper Company, recently donated $240,000 to the Nature Conservancy, with the objective of protecting the sustainability of the forests of Michigan's Upper Peninsula. This partnership program actually allows the nature conservancy to visit company forest lands, in order to achieve a greater understanding of biodiversity, and the necessary requirements for achieving truly sustainable forestry. According to Volkers, "You could argue that this has as much, if not more impact in protecting biodiversity, rather than trying to put protection contingencies in place for threatened lands." Volker also makes the point that the companies are responding not only to consumer pressures, but also to the desire to run their businesses in more efficient and sustainable ways.

Financial indices such as the Dow Jones Sustainability Index (DJSI) are designed to track the financial performance of groups of companies, based on their sustainability—in terms of economic, environmental, and social parameters. This type of index has already proven beneficial, since it essentially rewards the companies it tracks for their positive environmental efforts, actually rating them in terms of their levels of sustainability.

One example of a company that understands that being "Green" is also good for business is British Petroleum. BP has made a significant investment in solar energy, and recently received an award for sustainable development from the European Union for its role in developing a new process for removing sulfur from gasoline. In 1998 the company set a goal for itself to reduce greenhouse gas emissions by ten percent from its 1990 level, predicting that it could achieve this goal by 2010. Surprisingly, BP was able to meet its target by 2001—at no additional cost to the company. In the words of BP Senior Advisor Charles Nichols, "We're convinced there doesn't need to be such a harsh trade-off between economic growth fueled by energy consumption on one hand, and a clean environment on the other. People expect successful companies to take on challenges, and we have accepted this challenge."[106]

The Greening of Japan's Cities

With average summer temperatures on the rise, the cities of Tokyo and Fukuoka have discovered a simple, inexpensive, and natural way to counteract the heating effects of "Urban Heat Islands," created by vast expanses of concrete and paving materials. The Japanese Urban Development Commission (UDC) has initiated a program for "greening" Tokyo's rooftops; using rooftop grass as a cheap alternative to energy-hungry air conditioners.

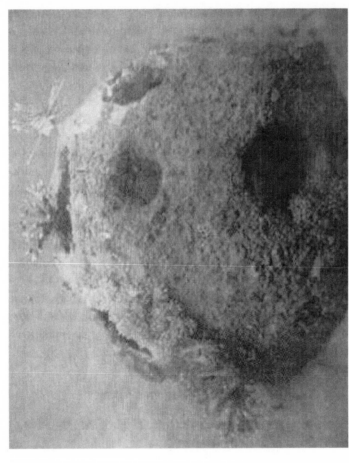

Reef Ball—A hollow concrete habitat for marine fishes and invertebrates, used to create artificial reefs. Courtesy of Reef Ball Development Group.

Solar Tower Generating Plant—A 1000-meter tall chimney/greenhouse structure combining solar power and wind power—to be constructed on the Australian Outback. Courtesy of Energen Global, Inc.

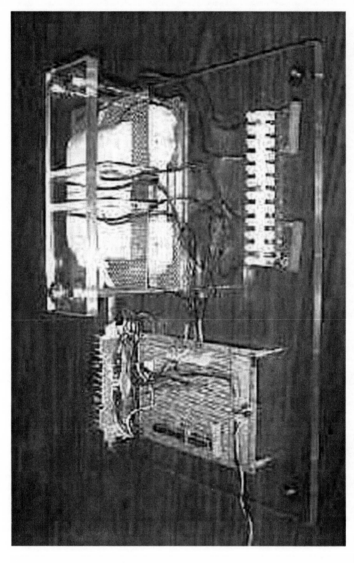

MEG Device (Motionless Electromagnetic Generator)—The first zero-point energy device ever to be granted a U.S. patent. Courtesy of Tom Bearden, Ph.D.

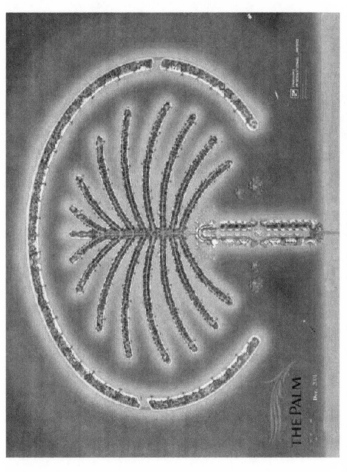

Palm Island—One of two colossal manmade Islands being constructed in Dubai, United Arab Emirates. Courtesy of the Government of Dubai.

153

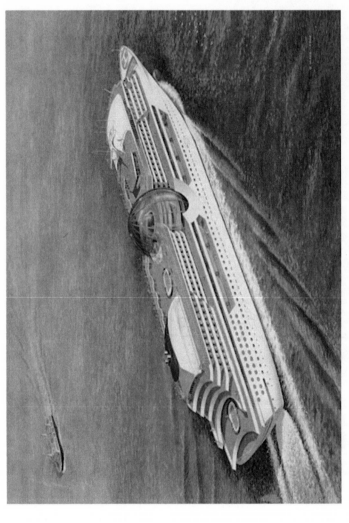

Gaiaship—A futuristic luxury liner, designed to be a floating center for global peace and cultural enrichment. Courtesy of Roar Bjerknes, founder and chairman of R B Media Pte. Ltd., The Gaiaship Project.

Lone Star City—Futuristic domed megalopolis, designed by Michels-Bollinger, Inc. Courtesy of Doug Michels, architect and Michels-Bollinger, Inc.

Circular City—Integrated city system of the future.
Courtesy of Jacque Fresco and Roxanne Meadows, The Venus Project

Mile-High Skyscrapers—Graceful tripod-shaped high-rise towers constructed of carbon fiber-reinforced pre-stressed concrete. Courtesy of Jacque Fresco and Roxanne Meadows. The Venus Project.

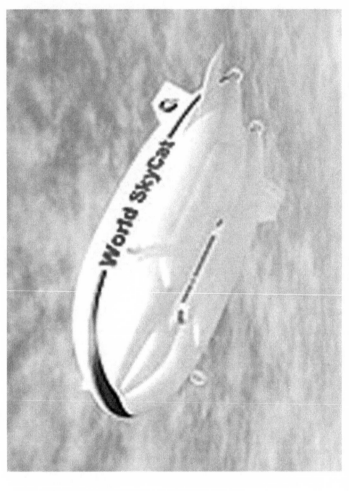

SkyCat 20 Airship—A unique hybrid airship, which combines a rigid aerodynamic lifting body with a stable catamaran design. Courtesy of Mike Crosby, SkyCat Technologies.

Aerobus—A futuristic mass-transit system.
Courtesy of Dennis Stallings, Aerobus International, Inc.

New Species of Giant Freshwater Shrimp—This photo was taken in the early 1970's by Dr. Shao Wen Ling in Taiwan. Smaller specimen is the commonly farmed Macrobrachium rosenbergii. Large specimen is a new species unknown to science.

160

Large bioreactor at the Kuwait Institute for Scientific Research, developed for creating single-cell protein from oil. Photo by the author.

The UDC discovered that simply adding a six-inch layer of soil, water-absorbing perlite, and grass to the roofs of city buildings, high rises, and parking garages would cut the heating effects of a concrete structure by as much as 37°C (99°F). In an initial testing program, the installation of rooftop grass replaced 2,500 rooftop air conditioners.

The rooftop vegetation—a combination of Korean Velvet Grass and Lily Turf—requires little water, and needs weeding only twice a year. Besides providing insulation, turf rooftops offer the advantages of improving energy efficiency, cleaning up air pollution, and restoring a natural ambiance to urban settings.[107]

Earth's "Natural Internet"

Mushroom and fungus cultures play a major role in the replenishment and remediation of the global biosphere. Fungal complexity is also a primary indicator of a healthy ecosystem. Our planetary land masses are covered with fungal mats: networks of living cells, which belong to a biological kingdom only marginally understood by the scientific community. These fungal mats constitute the largest biological entities on the planet, with some "Individuals" covering an area as large as 20,000 acres.

Mycelial "Waves" grow outward at up to 2 inches a day, influencing all of the biological systems on which they depend. When one species matures and dies, a variety of different fungal species quickly establish themselves. Every cubic inch of soil is host to thousands of fungal species. Of an estimated 6 million fungal species, only about 50,000 have been scientifically catalogued.

This vast, interconnected mycelial mantle has been called "Earth's Natural Internet," since it grows quickly in plant and animal detritus, and is instrumental in recycling essential elements. When natural disasters disturb the environment, these "Fungal Champions" come to the rescue by capturing and recycling organic debris. By increasing our awareness of this important biological kingdom, and its role in the repair and restoration of our global ecosystems, we will gain a better understanding of the overall complexity and interactions of this "Biological Web of Life." This little understood biological resource also offers untold promises and possibilities for new discoveries in the fields of Bio-remediation and Medicine.[108]

Australia's Catholic Church and Environmental Reform

Australia's powerful Catholic Church, which represents some 25 percent of the country's population, has indicated its intention to mobilize its congregation to support environmental issues. The church's major environmental concerns include the irresponsible clearing of natural landscape, logging in old-growth forests, and protecting the Great Barrier Reef.

In May of 2002, Catholic Earthcare Australia, an environmental initiative, backed by the Australian Catholic Bishop's Conference, was launched at a special mass at the Church of St. Francis of Assisi in Sidney. According to Archbishop John Bathersby, who holds the chair of the Bishop's Committee for Justice, Development, Ecology and Peace, the ultimate goal for Catholic Earthcare Australia is, "to mobilize Australia's five million Catholics to take decisive action to protect the natural world before it's too late." The initiative originated from a statement by Pope Paul II, who urged the Catholic Church to pay greater attention to protecting the environment. In his 1990 World Day of Peace address, the Pope declared, "The Ecological Crisis is a moral issue....respect for life and for the dignity of the human person extends also to the rest of creation."

On January of 2001 the Pope addressed the negative impacts of humans on the environment as follows: "Humanity has disappointed God's expectations. Man, especially in our time, has without hesitation devastated wooded plains and valleys, polluted waters, disfigured the Earth's habitat, made the air unbreathable, disturbed the hydro geological and atmospheric systems, turned luxuriant areas into deserts, and undertaken forms of unrestrained industrialization." He continued, "We must therefore encourage and support the ecological conversion which in recent years has made humanity more sensitive to the catastrophe to which it has been heading."

In response to the Pope's pleas for "Ecological Conversion," the Australian Bishop's Committee for Justice, Development, Ecology and Peace created a new organization: Catholic Earthcare Australia. It appointed an advisory committee, which included Christine Milne, former Parliamentary Leader of the Tasmanian Green Party; Aboriginal Leader Mick Dodson, a popular champion for social justice; and Franciscan Priest, Father Andrew Grave. One of the first products of the new Earthcare initiative has been to distribute a 20-minute environmental video called *The Garden Planet* to all Catholic Schools in Australia. The video focuses on the critical environmental issues, which are currently facing Australia. These issues include threats to both ecological and human health from pollution, the logging of old-growth forests, and the threat of global warming.

Although the Catholic Church has previously played significant roles in debates on economic and social policy, Earthcare Australia is the first national initiative on the environment, and a shift that has been well-received by environmentalists. According to Alec Marr, Campaign Director for the Australian Wilderness Society, "The fact that the archbishops in Australia are taking the call by the Pope for an ecological conversion seriously, is extremely heartening. We are delighted that the Catholic Church is taking a leadership role amongst religious leaders to put the environment fairly and squarely on the national agenda. Hopefully it will bring a moral perspective to the damage that is being done to the environment."

The Catholic Earthcare Initiative has already created a ripple effect throughout the global Catholic community. The new environmental educational video was screened at a recent meeting of European bishops in Italy, where a number of the participants expressed interest in obtaining copies for distribution in their own respective regions. It is felt that the new movement toward an "Ecological Conversion" will also be enthusiastically received in the Asian-Pacific region, where the Catholic Church also has a significant influence on the general population.[109]

The "Enhanced Greenhouse Effect"

The identification of an Enhanced Greenhouse Effect "The mother and father of all environmental issues" is spurring us on to rethink the long-term shape of the global economy."

~ Jonathon Porritt, 2000, *Playing Safe: Science and the Environment,* p. 74.

The Cost of Magazines in Terms of Trees

A new Internet-based program called "The Paper Wizard," offers magazine publishers and readers the opportunity to calculate the numbers of trees that are consumed by printing any U.S. magazine on recycled or non-recycled paper. It is estimated that less than five percent of U.S. magazines contain any recycled paper at all, and that the publication of these magazines is estimated to consume approximately 35 million trees a year!

According to Frank Locantore, Project Director of Co-Op America, and member of the Magazine PAPER Project Coalition, "The Paper Wizard presents in very stark terms the impact that each magazine has on forests." For example, sample calculations done by Magazine PAPER'S staff members for *Cosmopolitan Magazine,* suggest that the Magazine could save at least 32,858 of the trees it presently consumes each year by switching to recycled paper, which would provide qualities which are similar to virgin paper. *People Magazine,* which presently uses approximately 546,134 trees a year, could save 54,613 trees each year by making the switch to recycled paper. *National Geographic Magazine* already saves some 2,255 trees a year from a total of 505,819 trees, by merely using ten percent recycled paper for the magazine's cover. If it were able to use recycled paper for the entire magazine, it could save an additional 48,552 trees yearly.

According to Susan Kinsella, Executive Director of Conservatree, a coalition member of the PAPER Project, "The Wizard is a great tool for helping publishers understand the environmental impact of their publications, and the tremendous benefits that can be obtained by switching to recycled paper." The Paper Wizard has been designed to formulate results, which range from a close estimate to a sophisticated calculation, depending on the complexity of the data entered into the program. The Wizard program integrates the data entered for weight and grades of paper used in both a magazine's cover, and the inside pages. It also takes into account the size of the magazine, number of pages, circulation, and frequency of publication.

The Paper Wizard can also calculate the numbers of trees that could be saved by switching to recycled paper. For example, using paper which contains between 10 and 30 percent post-consumer recycled stock for both coated and non-coated stock, offers similar performance to virgin paper, and can often be obtained at competitive prices.

The Magazine PAPER Project Coalition includes non-profit organizations such as Coop America, Conservatree, and the Independent Press Association. The PAPER Project focuses on preserving the health of forests and communities, by encouraging magazine publishers to commit to sustainable environmental practices. The project helps publishers understand the environmental impacts of their companies, and assists them in shifting over to more environmentally appropriate industrial practices.[110]

Cleaning up the Planet

In 1989, Australian Yachtsman, Ian Kiernan made the decision to do something about the trash he observed nearly everywhere he had traveled on his extensive ocean voyages. He subsequently initiated a campaign which eventually resulted in 40,000 people getting together for a "Clean up Sydney Harbor Day." This Event proved to be so successful, that the "Day" became a "Weekend," and then "Several Weekends."

What began as a local effort, eventually expanded to a national, and eventually to a full-blown global clean-up campaign! During the past decade, the coordinated efforts of "Clean up Australia" have resulted in the restoration of pristine windswept beaches, and sparkling clean cities.

The end result of Kiernan's efforts, which began back in 1989, has been the formation of a United Nations affiliated organization called, "Clean up the World," which now has satellite organizations in 120 different countries. When Brigitte von Bulow, an Australian volunteer who coordinates clean-up teams in the U.S., first visited Disneyland in Florida, she was horrified by the amounts of trash thoughtlessly discarded in the area. "I could have cried from the dirt and garbage on the road—so much paper and cardboard boxes." Brigitte is in the process of establishing "Clean up the World-USA," as a coordinating group, which will help organize smaller regional teams. In the words of Brigitte, "We want the new generation to see the beauty of the world, not just the materialistic aspect."[111]

Restorative Aquaculture:
A New Strategy for Planetary Healing

Restorative aquaculture can be defined as the protection and enrichment of specific marine ecosystems, such as coastal mangrove and seagrass communities, clam, oyster, and mussel beds and coral reefs. The concept also extends to include our coastal and oceanic fishing grounds. Protection of the mangrove and seagrass beds is especially important, since these marine habitats function as nursery grounds for many species of juvenile fishes, shrimps and other invertebrates. These fragile coastal areas are being seriously threatened nearly everywhere in the world, by human activities such as dredging, development, and commercial aquaculture. Seagrass meadows and coastal mangrove forests, in addition to being rich biological habitats, also provide valuable storm damage protection to coastal shorelines, by disbursing the wave energies from major storms.

Coastal and offshore fishing areas have also incurred major damage from bottom trawling and commercial over-fishing. Coral reefs throughout the world have already lost most of their large fish species several decades ago due to spear fishing and destructive fishing methods such as dynamite fishing and the use of toxic chemicals to collect specimens for the global aquarium trade. Specialized fisheries for live fish restaurants, and Asian pharmaceutical markets also continue to take a steady toll. The indiscriminate collection of exotic shells and marine specimens has pushed some species to the brink of extinction, and unbalanced these fragile marine ecosystems on a worldwide scale.

Restorative aquaculture uses existing aquaculture technology to breed marine organisms in captivity—by growing out the young in protected intensive culture systems which provide optimal conditions for their growth and survival. Many marine species are then released back into the sea as juveniles, while other fishes and invertebrates can be raised to adulthood, and then released back into the environment. In protected marine areas, such adult specimens will have the opportunity to quickly begin producing large numbers of offspring, thus re-establishing critical breeding populations of their own species.

Environmentally sustainable new programs should be established to set up aquatic nurseries, with technical and financial assistance from governments and environmental organizations. Such programs would provide indigenous people with sustainable cottage industries that would enhance (rather than deplete) their local fisheries resources. Small-scale "Fish Farms," established to produce the most desirable exotic tropical fishes, shrimps, anemones, corals, and mollusks, could thus provide entirely new and sustainable enterprises for the local fishermen. In exchange for the financial and technical assistance to establish these local businesses, a set proportion of the species produced would be released back into the wild as an "Environmental Tithe" to re-stock the local marine ecosystems.

Selected pristine marine areas in every region of the world should be designated as marine parks, which would function as "Biological Banks" and "Protected Grow-Out Areas" in conjunction with the local aquaculture farms. These international marine parks would also function to "seed" nearby areas which could be opened to commercial and sports fishing; while the protected areas could be opened up to recreational scuba divers and snorkelers, thus bringing in eco-tourist dollars to support the local economies.

The key element for success with restorative aquaculture lies in the fact that all marine organisms produce a surplus of eggs to compensate for mortalities from predation and other normal losses in the natural environment. Restorative aquaculture has the potential to effectively boost natural survival rates by factors of 100 to 1000 times, or more—thus providing scientists with a "Magical Ingredient" for restoring of the marine environment.

Restoring the Global Biosphere
Through Ecological Enhancement Technology

"Ecological Enhancement," is a new and enlightened concept for restoring the Global Biosphere. In essence, it involves combining the latest scientific knowledge and technology with traditional biological wisdom, with the ultimate objective of enhancing the natural environment. The basic methodology, in its simplest form, involves the following series of actions: 1) First, the particular area of the environment under consideration must be granted "Legal Preservation Status." 2) Then, a program must be funded and implemented to protect the designated ecological area from indiscriminate resource-plundering and bio-piracy, so it can be allowed the opportunity to fully express its own unique ecological characteristics. 3) Once protected from negative intrusions and influences, such nature preserves would serve as "Biological Banks." They would constitute prototypical reference centers—depicting in form and function, archetypal models of what constitutes a mature ecosystem within a particular geographical location. Once the basic flora and fauna of such an area have been catalogued, a system of ecologically appropriate trails could be established for purposes of outdoor recreation, nature photography, environmental education, and research. These sustainable activities, combined with ecologically focused resorts, donations and usage fees, would help fund the maintenance, protection, and perpetuation of the nature preserve, so that it would eventually become a self-sustaining institution, and a legacy for present and future generations.

In selected cases, natural ecosystems can be enhanced by removing, and/or neutralizing negative ecological elements. For example, in the smaller wilderness parks, removing some of the deadwood and underbrush would serve to enhance the forest as a whole, creating a park-like ambiance such as was common in European feudal societies (i.e. "Royal Forests"). Although the majority of these 21st century "Environmental Forest-Parks" would be accessible to the general public, these areas would still constitute Balanced Ecosystems, which would be created in the best interests of the trees, plants, and animals—as well as the humans that would visit or inhabit these forested areas. Over time, and in gradual stages, an "Enhanced Environment" could be created. Such enhancement could, for example, be achieved by adding selected trace minerals and organic mulch (created by pulverizing the excess brush and deadwood). These actions would serve to boost the growth and vitality of desirable species of trees and other vegetation. As nearly all soil types are nutrient-deficient in one aspect or another, balancing out the soil would tend to create more robust and healthy vegetation, which would exhibit exceptional characteristics of vitality and resistance to disease. Crowded trees could be thinned out to allow each tree or shrub sufficient growing space. Conversely, diseased, genetically weak, or ecologically incompatible trees, shrubs, and plant species could be eliminated, while adding tree saplings, bushes, and flower plantings of desirable species. This would serve to synergistically complement each forest ecosystem, as dictated by its own unique factors such as soil type, climate, latitude, and altitude.

Enhanced Environments could be funded by governmental or

non-governmental organizations. They could also be cooperatively owned and managed by groups of private individuals. These natural preserves would be created and maintained mainly for purposes of outdoor recreation, environmental education, scientific study, and eco-tourist oriented activities; in conjunction with other forested areas, which could be designated as sustainable forest farms for the commercial production of timber resources. In the future, enhanced environmental forest parks could even be enclosed in huge "Eco-Domes." Such massive domed enclosures would allow for the creation of "Global Forest Parks," which could be integrated into the basic design of futuristic eco-cities. These eco-cities would synergistically combine the best of natural and urban living, using the natural vegetation for shade and to purify the atmosphere. In these enhanced environmental systems, the hardiest and most attractive species from around the world could be brought together with a corresponding selection of bushes, grasses, flowers, and wildlife. Light quality, day length, temperature cycles, and humidity, would also be controlled to create optimal conditions for growth and vitality.

A similar approach might also be taken with artificial reefs; creating "Enhanced Marine Preserves" in areas away from natural reefs. These enhanced coral gardens could be "seeded," and "weeded," using technology assisted nature to maintain an optimal balance of only those ecological elements which would be maximally appropriate for the health and diversity of each unique ecosystem.

The Greening of Peru's National Consciousness

In May 28 of 2001, a 5,225 square mile section of pristine Andean rainforest—slightly larger than the State of Connecticut became The world's newest national park. The Peruvian Government signed into law the Parque Nacional Cordillera Azul, one of the area's last extensive areas of virgin forest, which is still undeveloped and mostly uninhabited. Funding for the new program is being provided by outside agencies, and will go directly to park management.

Peru's President is very enthusiastic about the park, as are the local people who independently requested it. From past experience, they felt that environmental protection policies would take years to implement. This time, however, only two weeks elapsed between the President's announcement and the creation of the park, thus setting a new prototypical example of environmental action for the rest of the world.

The local people decided that creating a park would stop individuals from higher elevations from moving into the rainforests and clearing the forested areas for farming, as opposed to the idea of developing programs for sustainable forestry, and forest farming.

A team of scientists, who recently explored areas of the new National Rainforest Park, have already identified 28 new plant and animal species, and an entirely new species of armored catfish. By supporting the creation of their new National Park, the people of Peru have made a priceless gift to the Planet—and to the generations of people who follow.[112]

Trading Debt for Rainforest Protection:
A "Win-Win" Eco-Policy for Peru

In a precedent-setting agreement, conservation groups joined with the United States government to provide the necessary funding for a "Debt-for-Nature" swap—a win-win agreement, designed to relieve national debt in exchange for a commitment by a foreign government to invest a set percentage of its local currency in environmental conservation projects. Under the new agreement, The Nature Conservancy, Conservation International, and The World Wildlife Fund contributed a total of US $1.1 million, in exchange for a U.S. government allocation to cancel a portion of Peru's national debt to the United States. As a result of this agreement, Peru will save approximately $14 million in debt payments over the next 16 years and in exchange, will provide the local currency equivalent of about 10.6 million dollars for ecological conservation projects over the next 12 years.

The U.S. Treasury Department's Undersecretary for International Affairs, John Taylor, had this to say about the new agreement, "From the top of the Andes to the Amazon Basin, Peru is home to 84 of the 103 types of 'Life Zones' found on Earth—with nine different life zones existing in Macchu Picchu alone. The funds generated will go towards protecting rainforests in Peru, including the Peruvian Amazon. This area is home to dozens of endangered species, such as the jaguar, harpy eagle, the giant river otter, black caiman, several species of macaws, and rare plants such as walking palms and giant water lilies." The ecological debt swap will generate much needed funds for local conservation groups in Peru, who will use them to protect ten tropical rainforest areas within the Peruvian Amazon, which cover over 27.5 million acres—an area about equal to the state of Virginia. According to Peter Seligmann, Chairman and CEO of Conservation International, "This strong international partnership marks a key step in protecting a spectacular place that is among the biologically richest on Earth, and is now facing imminent threats."

Environmental threats to the newly protected area included habitat loss from the unsustainable logging of mahogany and cedar, conversion of forest lands to agriculture, mining, gas exploration; and unsustainable harvesting of tropical forest products such as hearts of palm and Brazil nuts. According to Steve McCormick, President of the Nature Conservancy, "We have long advocated the use of The Tropical Forest Conservation Act as a tool to help protect vital rainforests around the globe, so being able to pitch in to help make this agreement happen is particularly gratifying to me."

Peru's debt-for-nature swap means that one of the world's great "Biological Libraries" can be preserved for future generations. Says Kathryn Fuller, President of the World Wildlife Fund, "We applaud the Governments of Peru and the United States for helping protect these places of exceptional wonder and beauty." Peruvian conservation groups will be able to use the new funding sources for the establishment, protection, and maintenance of parks, protected areas, and wildlife preserves. They will be able to improve their National Resource Management Systems, and to train individuals and organizations in the basics of conservation. Funding can also be used for development and support of people living in or near these new tropical forest reserves in ways that will protect and preserve their natural ways of life. Some of the funds will also be used for the identification and research of medicinal plants.

The U.S./Peru Agreement is the second U.S. debt-for-nature swap, and the fifth agreement to be concluded under The Tropical Forest Conservation Act. Says U. S. Undersecretary Taylor, "We are still looking to extend the benefits of the programs, and have already seen agreements with Bangladesh, Belize, El Salvador, and Thailand. We expect to conclude an agreement with the Philippines this year."[113]

Brazil Inaugurates the World's Largest Tropical National Park

On August of 2002, a section of Amazon rainforest larger than the state of Maryland became the world's largest tropical national park—granting protection to a biological treasure trove of undescribed plant and animal species. In a decree, signed by President Fernando Henrique Cardoso, the new Tumucumaque Mountain National Park was created. The new park, which encompasses an area of 9.6 million acres, includes an uninhabited region of virgin rainforest, and vast expanses of forest-blanketed granite-topped mountains, rising to heights of up to 2,300 feet above the forest floor. In the words of President Cardoso, "Plants and animals that may be endangered elsewhere will continue to thrive in our forest forever."

The new Tumucumaque Park is 568,000 acres larger than the Slonga National Park in the Republic of Congo, which previously held the title of the world's largest tropical park. Native inhabitants of the new forest preserve include anteaters, giant armadillos, jaguars, sloths, harpy owls, and black spider monkeys. To date, scientists have identified eight different species of primates, 350 species of birds, and 37 different types of lizards. Environmental groups who were instrumental in creating the new park include The World Wide Fund for Nature, and Conservation International. Says, Brazil's Director for Conservation International, Roberto Cavalcanti, "The Park is very important because it helps consolidate the world's last roadless wilderness. Although much of the Amazon is still wild, there are roads running through it."

In many areas of the Amazon region, roads have unfortunately been a key factor in accelerating the destruction of virgin rainforest, since they provide access to loggers, prospectors, and settlers. Already, deforestation has destroyed over 15 percent of Brazil's Rainforest in the Amazon Basin—an area equivalent to about 1.35 million square miles. Since the Tumucumaque Park has many waterfalls, whitewater rapids, and rivers that are virtually impassible year-round, the area is one of the last few regions on Earth that remains unchanged by human activities. In the words of Jose Maria Cardoso da Silva, Director of Conservation International for the Amazon Region, "This Park today looks much like it would have hundreds of years ago, since Tumucumaque has never been deforested."

Environmental Secretary for Biodiversity and Forests, Jose Pedro de Olivera Costa, is hopeful that the millions of dollars in funding promised by the World Bank and Global Environmental Facility will help insure that Tumucumaque avoids the fates of other Amazon parks, where critical shortages of forest rangers and infrastructure have rendered these parks vulnerable to illegal logging and mining operations; yet unfortunately remaining inaccessible to eco-tourists and the general public. Says Costa, "We want Tumucumaque to be the first of a series of parks that includes visitors and ecotourism. We want to give it model treatment, everything we think is necessary for a park."

During the initial developmental phases, the park will be open only to scientists, who will establish a plan for how best to combine preservation of the national environment with ecotourism. In the words of Garo Batmanian, CEO of the World Wide Fund for Nature, "Because most of the land in the Amazon is still in the government's hands, the environment can still have a vision for zoning the Amazon."[114]

Australia Creates Earth's Largest Marine Sanctuary

On October of 2002, the Australian government officially announced the establishment of the world's largest marine sanctuary. The New Heard Island and McDonald Islands Marine Reserve is located some 2,796 miles (4,500 km) southwest of the Australian Mainland, and 620 miles (1,000 km) north of Antarctica. This newly protected area is one of the most isolated places on Earth. It includes an active volcano, covered with snow and glacial ice, which rises abruptly from the stormy waters. The New Marine Reserve encompasses an area of 25,096 square miles (an area about the size of Ireland), and is one of the most pristine environments ever to be granted protection from commercial development.

Australia's Environment Minister, Dr. Donald Kemp, announced that the New Reserve was created to protect the habitats and food resources for some of the world's most unique sea creatures, including the southern elephant seal, the Sub-Antarctic fur seal, and several species of penguins. In Klemp's own words, "The Heard Island and McDonald Islands Marine Reserve is being declared to protect the conservation values of the region in an integrated and ecologically sustainable manner."

The new ecological preserve is truly unique, since it is the only Sub-Antarctic island group that contains no foreign species introduced by humans from other areas. This makes the region ideal for scientists to study an undisturbed set of interrelated ecosystems, which include terrestrial, freshwater, coastal, and marine ecosystems, where the plants and animals have all evolved in authentic natural settings.

The beaches in the new reserve are home to vast colonies of seals and penguins, and provide some of the most unspoiled and magnificent wildlife vistas on Earth. For example, the area is home to the world's largest colonies of macaroni penguins, whose numbers alone are estimated to reach two million. The region also hosts two species of albatross—the light-mantled sooty albatross, and the black-browed albatross. Benthic marine environments in the park include among their myriad inhabitants a variety of soft corals, glass sponges, and giant barnacles.

Australia's only active volcano, "Big Ben," is a 9,006 foot (2,745 m) peak, located on Heard Island. The island's precipitous shoreline is so remote and rugged, that only two successful land-ings have been recorded since the island's discovery over a century ago. Because of their spectacular beauty and remote wilderness, both Heard and McDonald Islands were placed on UNESCO World Heritage List in 1997.

World Wildlife Fund Australia has nominated the New Marine Reserve for Australia's highest conservation honor, "A Gift to the Earth." According to Senior Marine Policy Officer for WWF-Australia, Margaret Moore, "This is one of the most significant conservation decisions taken by Australia. It acknowledges that we have exercised our sovereignty in the region with governmental response, and the declaration of this major new marine resource, one of the largest most fully protected marine areas on the planet, second only to another Australian reserve in the same region, maintains that great tradition."[115]

Canada Plans Ten Huge National Parks

In a move to protect unique natural landscapes and wildlife, the Canadian government has announced plans to create ten new national parks, and five new marine conservation areas over the next five years. The new parks will cover an area of approximately 39,000 square miles—nearly doubling the area of the presently existing Canadian National Parks.

In a speech, announcing the decision to create the new National Parks, Prime Minister Jean Chretien stated, "Canada is blessed with exceptional national treasures. We owe it to Canadians and the world to be wise stewards of these lands and waters." Chretien also stated that the government would do more to improve the existing national park system, which has often been criticized during recent years for lack of adequate funding.

According to Canadian Heritage Minister, Sheila Copps, the new National Park Expansion Plan is the most ambitious plan to expand the existing park system since the first National Park was established in Banff, Alberta in 1885. Said Copps, "National Parks are in our hearts. They are important to our identity as Canadians. With this far-reaching plan, we are fulfilling a Canadian dream."

Sites for the seven new national parks which already have been identified include: 1) The Gulf Islands in Western British Columbia, one of the nation's most biologically diverse and threatened natural areas; 2) Ukkusiksalik, located in the vast Arctic territory of Nunavut, a wilderness area which is host to polar bears, caribou, peregrine falcons, and musk oxen; 3) The Torngat Mountains in the remote Arctic wastelands of Labrador, noted for its spectacular mountains, fiords, polar bears, and caribou herds; 4) The Mealy Mountains in Labrador, with its boreal forests, extensive bog lands, and wildness rivers; 5) The Lowland Forests in the Province of Manitoba, which features the nation's longest sand spit, as well as fresh-water marshes and bat caves; 6) Bathurst Island in the Territory of Nunavut—home to vast herds of Peary caribou; and 7) The Eastern Arm of Great Slave Lake in the Northwest Territories, which features spectacular cliffs, as well as abundant moose, bears, and wolves. The three remaining wildness park sites are still in the process of being defined by Canadian National Park Authorities.

The new marine conservation areas will be created in British Columbia's Gwaii Islands in the western region of Lake Superior, and in the southern area of the Strait of Georgia in British Columbia. The two remaining marine conservation areas have yet to be identified.

Premier Chretien's announcement was generally well-received, but the Canadian Parks and Wilderness Society (a non-profit organization) expressed its concern that the Canadian government needed to increase funding for the National Parks System. According to the Society's Vice President, Harvey Locke, "It is clear that parks will not be established, nor will the declining health of existing national parks be addressed, without money committed to the cause." Earlier, an independent panel appointed by Heritage Minister Cobbs, had recommended that US $206 million be invested over a five-year period—specifically for the purpose of restoring the ecological integrity of the existing National Parks System.[116]

Russia's Lake Baikal and Siberian Parks Endangered

There are few natural wilderness spots on Earth as beautiful as Lake Baikal. In the past, when the Soviet Union prospered, it was a popular destination for summer tourists, who wished to experience the beauty and tranquility of Russia's natural wilderness. Today, it has become a lesser-known wilderness destination that eco-tourists are on the verge of rediscovering. Adventure travelers might wish to visit this pristine area soon, however, since in the next decade there may be little left to rediscover.

A group of Russian oil companies, along with British Petroleum, currently plan to build a series of pipelines that will skirt the Lake Baikal region to feed the expanding Chinese demand for oil. The rich oil deposits will *not* go to enrich the local economy, however, as Russia is simply too much in need of hard currency. Accordingly, the big oil companies in Moscow and elsewhere find it more profitable to sell the oil to China, despite the fact that this means building a 2,000-mile pipeline, which will run through several national parks and pristine wilderness areas along the way.

Local environmental activists are already launching plans to fight the proposed pipeline. Their primary objective is not only to protect the environment and their predominantly Tibetan Buddhist culture; they are also worried about the economic changes that the pipeline will bring to the region. They are concerned with maintaining natural productivity of the region, and with supporting the new eco-tourism industry.

The shores of Lake Baikal are home to three national parks, each the size of Yellowstone, as well as four other equally large wilderness preserves. In addition, there are also hundreds of Buddhist temples, which have been painstakingly rebuilt over the past decade.

Recently, several international organizations have come together to recognize the lasting and sustainable benefits that eco-tourism can bring to the Lake Baikal Region. For example, the World Bank estimates that, with new ecologically appropriate policies and government support, eco-tourism could easily become the Lake Baikal region's major industry. The U.S. Agency for International Development, in cooperation with the Foundation for Russian American Cooperation, has made the decision to fund the Baikal Federation of Eco-tourism.

Unfortunately, the most desirable eco-tourism resources are also those parks and wildness areas that are the most likely to be threatened by the proposed pipeline. One such park is located in the Tunka region, to the southwest of Lake Baikal. This magnificent wilderness vastness is characterized by sweeping forested valleys, which are framed by towering mountain peaks. It includes dozens of mineral springs, and whitewater rivers, which are ideal for rafting. Park officials seem convinced that the Tunka Park will attract eco-tourists and wilderness adventurers, as well as religious pilgrims. Their main concern, however, is that the special sense of "wilderness attraction" will be negatively impacted by the ugly scars of oil pipelines, running through mountain passes, over valleys, across the rivers and springs—and on through into China.

The oil companies that are pushing for the pipeline claim that at least 30 percent of pipeline-associated employment will go to benefit the locals—especially the poorest sector. What they fail to mention is, that several years ago, on Sakhalin Island, oil companies made similar promises to locals in the area. However, not even 10 percent of the project ended up using local workers. Since there is no mechanism currently in place for local governments to collect taxes on the oil, virtually all revenues and fees are likely to revert back to the central government in Moscow.

Recently, however, many more tourists have been visiting the Baikal region in search of wilderness adventure. With the creation of the Great Baikal Trail, plus the recent growth of the Baikal Federation of Eco-tourism, there is good reason to hope for the creation of long-term service-oriented jobs, which will center on sustainable eco-tourism. The Great Baikal Trail, presently being developed by a joint effort between the Baikal Foundation of Ecotourism and the national parks at Baikal, will eventually skirt the entire circumference of the Lake— some 1,000 miles. With the help and support of Earth Island Institute, the Tahoe Rim Trail, and the Foundation for Russian American Economic Cooperation, the Federation plans to host an active program of volunteer trail-building during the summer months.[117]

Basic Strategies for a "Global Green Revolution"

1) A set of global guidelines and economic subsidies should be created to assist industries and institutions in becoming more energy-efficient and non-polluting, while increasing their operational efficiency and profit margins.

2) New, effective strategies for corporate cooperation should be established to insure maximally efficient use of energy, raw materials, by-products, and finished goods. Cooperative and mutually productive relationships should also be established in areas of packaging, marketing, shipping, and waste-stream processing technologies.

3) Strict international guidelines should be established for mining and harvesting Earth's natural resources. These guidelines would embody global treaties, environmental incentives and penalties, international military enforcement of regulations, and a consumer-driven transformation of the global marketplace through investment in, and purchase of, approved "Green" goods and services.

4) An international plan should be put in place to protect and monitor the global commons. This plan would be designed to preserve our oceanic, atmospheric, and terrestrial natural resources. Environmental satellite surveillance programs could be coordinated with "Green Peacekeeping Forces" to protect the global commons, and wildlife sanctuaries around the world.

5) Economics would be acknowledged as the primary driver for mega-transformation into new sustainable lifestyles for the global community. Included within this concept would be educational eco-tourism, the creation of cooperative wilderness parks, a global marine and coral-reef park network, resort-focused space tourism, and "Virtual Safaris," into wilderness preserves, ocean depths, and outer space—all of which could be conveniently accessed by anyone, anywhere on the Planet.

METASTRATEGY VI

NEW GLOBAL PROGRAMS FOR WASTE REDUCTION, TREATMENT, & RECYCLING

**The attachment to the material world
as our primary source of happiness
lies at the root of much of the craziness
that humanity perpetuates upon the world.
It is this that leads us
to consume resources we do not need,
to treat other people as elements in an equation,
to discharge our refuse out of sight,
and to mistrust and abuse our own bodies.**

~ Peter Russell—*The Global Brain Awakens*, p. 24.

BioWaste as a Renewable Resource

Animal waste is one of the major causes of water pollution on the Planet. Every day, the animals destined for the dinner tables of America produce about 20 billion pounds of waste. This amounts to 250,000 pounds per second! A typical U.S. pig farm produces an amount of waste equivalent to a city of 12,000 people, while all the livestock in America combined, produces twenty times more excrement than the entire human population![118]

As world population continues to exert its relentless pressures on the planetary ecosystems, existing landfills are approaching capacity. These landfills contribute to atmospheric and water pollution, by leaching their toxic materials into the environment. Modern agricultural technology—with its heavy reliance on chemical fertilizers and pesticides—is another major contributor to surface and groundwater pollution.

Canada-based International Bio-Recovery Corporation, of North Vancouver, B.C., has developed a proprietary new technology, which offers a practical solution to the problems associated with global bio-waste pollution. The company's unique thermophylic bacterial process converts biodegradable human and animal wastes into pathogen-free, nutrient-rich organic fertilizers. This new technology accelerates the natural biodegradation process from a period of several years to 72 hours! This exciting new technology thus provides a viable alternative to expensive, and environmentally inappropriate landfills, and waste incinerators. Revenue is generated from the collection and processing of wastes, as well as from the sale of organic fertilizer products.[119]

In the United States,
the amount of waste generated by Livestock
is *130 times* that produced by Humans.
A single Hog Farm,
which recently began operation in Utah,
will soon produce more waste
than all the people in Los Angeles.

~ Ed Ayres, 1999—*God's Last Offer:*
Negotiating for a Sustainable Future, p. 277.

Built Like a Brick S*** House!

A research team from I-Shou University in Taiwan has discovered that it is possible to incorporate municipal sewage sludge into ordinary building bricks. Since most countries in the world are running out of landfill space, the new process promises to ameliorate a major waste disposal problem, while providing a basic construction material that does not involve the depletion of Earth's natural resources.

According to Research Team Leader, Chih-Huang Weng, "It's a win-win situation, because it converts the waste into useful materials, and alleviates disposal problems." The concept evolved because of the need to find new ways to dispose of the 670,000 tons of sewage sludge produced yearly by the island of Taiwan.

Although the solid material Weng's research team used for the project was filtered out from industrial effluent, he sees no reason why domestic sewage would not serve equally well. Optimal results were achieved when ten percent sewage sludge was added to the clay for making the bricks, however the process seemed equally effective when up to 30 percent of the sewage sludge was added to the so-called "Biobricks."

In addition to disposing of the sewage sludge, the new process provides other benefits. When the Biobricks are fired in the kiln at 900°C, not only are potentially harmful microbes killed, but toxic heavy metals are also locked up in the process. The finished bricks have no odor at all. Wang understands that the public may need a little coaxing before the new process is completely accepted, and comments that, "Legal approval and public acceptance still remain to be sought." It is also possible that the new Biobricks may need to meet special standards, to make sure they do not pose any threat to human health.[120]

Transforming Chicken Manure into Fuel

West Virginia chemical engineer, Al Stiller, has discovered a new process for producing fuel from liquefied chicken manure. A mixture of 35% liquid waste and 65% diesel fuel performs as well as pure diesel fuel. The process has important implications for reducing America's dependence on foreign oil, while solving major waste and water pollution problems for the poultry industry.

Says Stiller, "Even in a small state like West Virginia, Poultry Farming is a $200 billion business." The average poultry farmer has about 500 tons of waste to dispose of every year. Converting this amount of waste to fuel would create an environmental "win-win" solution, and also generate a cash value of approximately a quarter of a million dollars.

Horse and cow manure may also prove to be suitable as a fuel blend, but since poultry farmers were facing the most serious environmental pollution problem, the research was focused in that particular area. Practical applications for the residue that remains after the liquefying process are also currently being investigated.[121]

Four-Legged Pachyderm Paper Factories

Researchers at Thailand's Elephant Conservation Center have discovered a novel way to dispose of large amounts of elephant dung—by transforming it into clean, odorless, decorative paper. The founder of a new company, called Maximus, Ltd., stumbled on the idea when he read about a Kenyan game ranger, who had developed a way to transform elephant waste into paper.

Biological speaking, an elephant is truly a "Walking Paper Factory." To create paper from elephant dung, the dung is first liquefied with water, and strained through a sieve to remove the coarse fibers. The residue is then formed into thin sheets, using simple hand-made paper techniques, and then dried in the sun. The decorative paper thus produced can be used for a number of artistic and commercial purposes, and the residue from the process can be used as organic fertilizer.

In terms of ecological sustainability, this innovative waste conversion concept serves to produce a useful product, it generates jobs and income for the locals, and has helped solve the problem of disposing of the 4,400 pounds of elephant waste produced daily by the 40 elephants at the Thailand Conservation Center.[122]

PaperCrete and Paper Adobe

PaperCrete is industrial-strength paper mache, created from a mixture of paper, cardboard, sand, and Portland cement. The dry ingredients are first mixed with water into a slurry, and then cast into blocks or panels, which are dried in the sun. Since PaperCrete is 90 percent air, it is light in weight, and thus functions as an excellent insulator. It is also remarkably strong. Because all the ingredients are free waste materials (except for the sand and cement), PaperCrete is an amazingly inexpensive and ecologically appropriate building material.

Paper Adobe is a mixture of dirt, paper, and water, which can be used to create durable building blocks that are lighter in weight, and have a significantly higher insulating value than conventional adobe. By using Paper Adobe in place of conventional building materials, costs for building a home can be reduced to as little as 16 cents per square foot.[123]

Artificial Reef Memorials

Due to the high costs of burial plots, and the ecological impropriety of land-intensive cemeteries, the practice of cremation has been growing dramatically in North America. By 2010, this procedure is expected to take place in approximately 40 percent of all funeral services.

Georgia-based Eternal Reefs, Inc. is a company that has gone beyond the basic concept of scattering the ashes of the dead over the surface of the sea. It has created permanent living memorials in the form of artificial reefs. Eternal Reefs, Inc. mixes the ashes of the dearly departed into an environmentally approved concrete. The liquid mix is then poured into fiberglass molds, which form hollow, flat-bottomed concrete spheres which are punctuated by a Swiss cheese pattern of holes. These structures, which can be created in several different sizes, are then clustered together in environmentally appropriate locations on the ocean floor. Within a few months, they become encrusted with corals and other marine organisms, and are thus transformed into marine habitats for a variety of fishes and other marine creatures (http://www.eternalreefs.com).

To date, more than a hundred eternal reefs have been established along the coasts of North Carolina, South Carolina, and Florida. Prices for an individual memorial reef sphere range from $850 for a spot in a 100-person "Community Reef," to $3,200 for the "Individual" 4,000-pound Atlantis model.[124]

Cleaning Up Water Pollution with
Bioremediation Technology

The Government of the U.K. has launched a US $24 million research program, aimed at reversing damage to its polluted waters. This new program employs a process called Bioremediation Technology. The program will be funded equally by government and private industry. If all goes according to plan, microorganisms could soon be cleaning up nearly half a million acres of contaminated industrial sites in the U.K. In this process, bacteria are used to transform toxic substances into simpler, less toxic products.

Certain bacteria have the ability to break down the hydrocarbons in oil and gas spills, while others, called phytoremediators, can even absorb radioactive elements and heavy metals from industrial waste sites. One species of bacteria has even been discovered that resists the effects of ionizing radiation by repairing its own damaged DNA. In the United States, scientists have engineered a unique microbe that "eats" waste from nuclear sites, transforming it into less harmful materials. This process has widespread implications for cleaning up nuclear waste dumps on a worldwide basis—a problem which has an estimated price tag of $300 billion in the U. S. alone.[125]

Hydrogen Gas from Wastewater

Nevada based Hydro-Environmental Resources, Inc. has developed a new alternative energy system that converts ordinary water (even wastewater) into hydrogen gas, plus electrical power and clean water. According to company spokesmen this revolutionary alternative energy system is self-contained, efficient, and cost-effective. Although the process requires a patented additive, it is said to function without an outside energy source.

After many years of research and development, the company recently revealed its new technology, and claims to have a fully functional working model. The firm is presently working to upgrade the technology to create medium-sized power plants, which would be ideally suited for clean energy communities. It also hopes to eventually build larger power plants on an industrial scale.[126]

From Sewage to Wattage

A waste water treatment plant in Washington state has developed a new technology to capture methane gas from sewage, and feed it through a fuel cell to produce as much as one megawatt of electrical power. The $18.8 million plant—powered by a fuel cell energy system—has a fuel-to-electricity efficiency of nearly 60 percent!

Using this plant as a prototypical model, if 500 similar wastewater treatment plants throughout the United States could be converted to this process, the potential exists for generating approximately 500 megawatts of electrical power.[127]

Japan is Tops in Global Can Recycling

In the year 2000, over 80 percent of the aluminum cans purchased in Japan were recycled. In comparison, about 60 percent of aluminum cans were recycled in the United States and 43 percent in Europe for the same time period.

Of the cans recycled in Japan, nearly 176,000 tons (74.5%) were used to make new aluminum cans, with the rest being used for die casting and other processes. According to the Japan Aluminum Can Recycling Association, since aluminum cans are the most expensive beverage containers, they tend to be the most recycled items in the country. In Japan, reusing aluminum containers saves 97 percent of the energy used, than when aluminum is manufactured directly from raw materials.[128]

Ecological Packaging Perspectives

Whether "Paper" or "Plastic," discarded packaging material has become a significant feature of the global landscape.

Approximately one-third of the weight and about half the volume of America's solid waste consists of packaging materials. This equates to approximately 300 pounds per person per year, but does not include the 400 million wooden pallets that are used only once or twice, and then dumped into landfills—enough wood to frame about 300,000 houses.[129]

A recent study by the grassroots recycling network estimates that between 1990 and 1997, plastic packaging increased five times faster by weight, than the amount that was recycled.[130] Add to this the 125 million take-out cups used *daily* by U.S. coffee drinkers, and the incredible magnitude of this problem becomes glaringly apparent! Some cities in China have already decided to put a stop to this mindless waste of resources, by declaring outright bans on Styrofoam fast-food containers. (Half a million of these containers are used *every day* in Shanghai, alone!) Another major packaging-related waste problem, which needs to be approached with a combination of recycling, and using biodegradable materials, is the incredible volume of the 19 billion polystyrene packing peanuts, which are estimated to be used every year in America![131]

What really complicates the global packaging scenario is the underlying fact that a vast array of packaging and packing materials are produced in virtually every country in the world—using different technologies and materials. In most cases, uniform packaging standards simply do not exist! If we are to comprehend, and begin to resolve this complex and multifaceted environmental problem, we must first begin to understand that packaging does much more than just facilitate global commerce, it also changes the way we actually live our daily lives. A significant amount of packaging exists simply for advertising and sales purposes. In other words, packaging is closely linked to consumer tastes and demands. "Convenience Packaging" saves us valuable time in food preparation. For example, average food preparation times have dropped from 2.5 hours per meal in the 1930's, to 15 minutes per meal in the 1990's!

New packaging technologies have also facilitated the shipment of food products for thousands of miles from producer to table. For example, the post-World War II phenomenon of "Single-Use Consumer Packaging," has contributed to the fact that the food products now travel an average of 1500 miles from source to consumer. As explained by author and eco-journalist Daniel Imhoíf, the upside of this is that, "Thanks to miracles of modern packaging technologies, shelf lives of highly perishable meats, seafoods, and produce have been greatly extended. High-acid drinks and food can be stored and transported long distances. But this is food of a certain kind—transported long distances, highly processed, frequently irradiated, sterilized, and wrapped in light-weight plastics."[132]

In the year 2001 alone, American consumers discarded an estimated 20 billion eating utensils and 15 billion disposable plates. To help reduce the growing mountains of disposable trash, several innovative new technologies have been developed to produce biodegradable packaging, such as starch-based packing peanuts and biodegradable plastic bags. Although at first glance such "Bioplastics" and related packaging materials would appear to be ecologically appropriate alternatives to non-biodegradable plastics, many of these new materials require vast amounts of feedstocks from other sources. This alternative might thus tend perpetuate such crops as corn and soybeans—segments of the agricultural establishment which, in their present form, actually represent "Agents of Ecological Disturbance and Non-Sustainability" for the Planet.[133] Thus, what is really needed is a coordinated shift in producer and consumer habits and prac-tices, so that new eco-friendly, cost-effective packaging can be combined with ecologically sensible consumer lifestyles. This would serve to eliminate the vast majority of unnecessary consumer packaging, without compromising either the product or the consumer.

Having Your Plate and Eating It Too

In 2001 alone, Americans discarded over 60 billion cups, 20 billion plastic eating utensils, and 15 billion disposable plates. In response to this massive global assault on our landfills, innovative individuals in various parts of the world are working to create "Edible" food containers. For example, in Germany, for over a decade, edible wafer/waffle cups, bowls, and sheets have been used to serve french fries, sausages, cooked mushrooms, pizza, crepes, desserts, and ice cream. The main problem with this material is that it is breakable, has a low tolerance for fluids, and is more expensive than paper or plastic plates.[134]

Jerzy Wysocky, a Polish miller, has developed a technique for using organic wheat bran to make disposable dinnerware, which is both functional, and ecologically friendly to the environment. According to Wysocky, mill operators have long been searching for a productive use for wheat bran—a by-product of the grain milling process, and so cheap that, "Sometimes you can't even give it away." Several years ago, Wysocky patented a new compression device that transforms wheat bran into a variety of dinner plates, bowls, cups, and platters.

The edible dishes are presently available in a variety of sizes, but are all colored a deep brown—similar to the color of whole wheat bread. The tableware is entirely "organic" in that no additives or preservatives are used in the process. When discarded into a conventional landfill, the eco-friendly tableware will dissolve into organic compost within a matter of a few weeks. These food containers provide a good example of a "Win-Win" environmental product for the following reasons: 1) They are created from an organic waste product, 2) They effectively eliminate the labor and water used in dishwashing, 3) When introduced back into the natural environment, they rapidly degrade back into organic compost, and 4) If eaten, they bring a whole new meaning to the term "Regular Customer."[135]

Taiwanese entrepreneur Chen Liang-Erh has developed a way to make edible plates and bowls from oatmeal. He has formed a company which plans to produce six million plates and bowls a day during its first year and a half of operation. Within the following three years, the company expects to expand its operation to produce 14 million edible food containers per day.[136]

Potatopak, a UK-based Company, has developed a line of trays, plates, bowls, and cups made from Potato Starch. Surprisingly, these items have actually proved to be stronger, and have superior insulating qualities than even conventional disposable dinnerware. To reduce costs whenever possible, the company uses waste material from potato chip plants.[137]

An Italian Company called Novamont, has developed a line of eating utensils made from Maize, and an ice cream container made from Cookie Dough. These products are already in use at McDonald's outlets in Austria, Germany, and Sweden.[138]

In the United States, Edibowls of Santa Ana, California, produces a line of wheat-based, "Crispy Bowls," which they market to cafeterias and institutional customers.[139] Another U.S. company, Earth Shell Corporation, has undertaken a mission to transform the ecologically inappropriate PVC fast-food "Clamshell" into an environmentally friendly product...the "Earth Shell." Earth shells are created by combining potato discards or wheat gluten with limestone and special fillers, to produce a product which is biodegradable. The Company is currently partnering with Dupont, to create "Greener" alternatives for McDonalds and similar fast-food chains.[140]

The concept of edible dinnerware is not exactly new, since over ten years ago, Iowa State University Professor, Jay-Lin Jane, developed a process for creating edible dinnerware from corn, soy protein, water, and several other ingredients. Although the process has yet to be commercialized, it is ecologically friendly, and according to Professor Jane, "It will biodegrade in ninety days, and has a roasted nut flavor.[141]

Recyclable Diapers

In Holland, it has been determined that every baby uses an average of 30,000 diapers by the time it reaches the age of two and a half years. For the entire country, this creates approximately two million tons of waste a year! In addition to the massive amount of waste, disposable diapers occupy considerable space in municipal landfills. Worse yet, they take *many years* to decompose.

Now, parents in the city of Arnheim can drop plastic bags of disposable diapers into special bins, set up outside childcare centers. A local recycling firm has been contracted to extract wood pulp and plastic from the waste. This company also treats its wastewater, to transform the sludge that remains into marketable compost.

The Recycling Company collects the diapers for its recycling plant for the same price that was formerly paid to incinerate the diapers. As soon as the project becomes operational, the firm will collect about 440,000 pounds of diapers each week from bins, which are placed in convenient locations around the city.[142]

Biodegradable Plastic Bags

Plastic bags and bottles have become a major source of pollution nearly everywhere on the planet. Besides being very slow to degrade, they are responsible for countless deaths of turtles and other marine species that ingest them inadvertently—mistaking them for food. Plastic bags, bottles, and other plastic debris can be found floating in even the most remote tropical islands, or the far reaches of Antarctica.

The International Biodegradable Products Institute recently granted the very first "Compostable" Logo to a company called Biocorp, Inc., for its role in developing the first biodegradable plastic bag. The Compostable Logo represents a joint effort between the U. S. Composting Council and the Biodegradable Products Institute, to identify and certify products which are designed to biodegrade in municipal or commercial landfills, or composting systems.[143]

The "EcoSense" of Using Biodegradable Plastics

An entirely new class of biodegradable composites, manufactured from soybeans and other plant fibers, has been developed at Cornell University. According to fiber scientist, Anil Netravali,

"These new fully biodegradable, environmentally friendly, green composites have properties which in the future could replace plastic parts in the interiors of cars and trains, in computers, and in packaging materials and other consumer products. Although the plant-based fibers may not be as strong as graphite or Kevlar, for example, they are low in cost, biodegradable and replenishable on a yearly basis."

At present, most non-biodegradable plastics, which are made from petroleum- or synthetic-based polyurethane, polyethylene, or polypropylene, end up in landfills. Little of this material—which continues to steadily accumulate on a massive global scale—is suitable for either reuse, or recycling. plant-based "Green Composites" could thus conceivably become the environmentally appropriate and inexpensive alternatives for many mass-produced plastic products.

Researcher Netravali has been instrumental in developing new green composites made from ramie fibers (derived from an Asian perennial shrub), and from resins, which are derived from a soy protein isolate polymer. His research group continues to work with several other fiber sources, including kenaf, industrial hemp, pineapple fibers, and banana stems. According to Netravali, these new biodegradable plastics "will be made from yearly renewable agricultural sources, and would be environmentally friendly, because they would naturally biodegrade once they were thrown on a compost pile."

Other resin materials presently being considered for use in biodegradable plastics include: commercial resins like polyvinyl alcohol and polylactones, and microorganism-derived resins, which offer many exciting possibilities for the future of organic biodegradable plastic production on a commercial scale.

Biologically derived composites also offer new possibilities for providing "ecological relief" for The global landfill problem. Tree-free wood substitutes would also seem to be ideal for packing crates and building studs. Says Netravali, "Trees take 25 years to grow. The fibers we use, however, come from plants that grow to maturity in a year."[144]

Transforming Recycled Plastic into Bridges

The First Vehicular Bridge—built of unreinforced I-beams and components, made from recycled plastic—is now in operation. The single-lane fire equipment access bridge, which crosses a river in Wharton State Park, New Jersey is 42 feet long, and is strong enough to support the weight of a 36,000 pound fire truck.

The plastic bridge is constructed from a unique new composite polymer, developed at Rutgers University. The new material is composed of high-density polyethylene and polystyrene, derived from Recycled Packaging Materials such as plastic cups and milk containers. The new technology, developed and patented at Rutgers Center for Advanced Materials, uses a process called Immiscible Polymer Processing, which produces a composite material of great structural strength. Two or more plastic polymers are first melted together, and then extruded to form a fine microstructural matrix. The unique properties of the new material are the result of what is termed "an Oriented Microstructure," a chemical phenomenon that endows the material with "unexpected mechanical properties." This material is strong enough to be suitable for bridge I-beams, railroad ties, boardwalks, decks, docks, and other similar applications.

The Wharton Project represents the first demonstration of this new bridge-building technology. The bridge, which was completed in November of 2002, is impervious to water, and the effects of weathering. It is virtually indestructible, and requires none of the painting or maintenance normally associated with similar steel or wood structures. Exposure to sunlight and the natural elements causes a thin protective coating to form on the external surfaces of the structure, giving it a finish that tends to blend with the surrounding natural environment.

The bridge was designed by McLaren Engineering, a civil engineering design and consulting firm. Its basic construction consists of large I-beams, supported by vertical posts, with a series of smaller I-beams spanning the gaps between the larger beams. The upper surface is composed of tongue-and-groove decking planks, three inches thick. Materials for the bridge were fabricated by the Polywood Corporation, while the actual construction process was accomplished by a special engineering team, headed by Rutgers University professors Thomas Nosker and Richard Renfree.

The Rutgers research team is presently working to develop a variety of different materials, using technologies similar to those used to create the bridge beams. Among these new products are included; structural materials for the automotive and aerospace industries, specialized membranes and catalyst supports for the chemical industry, and unique new biomedical materials that, when implanted in the human body, initiate bone growth by promoting tissue infiltration throughout the material's porous polymer structure.[145]

Industrial Ecology and Co-Generation

Industrial ecology integrates industrial systems together similar to the way in which ecosystems interact in the natural world. In such a system, one organism's waste products become another organism's building material, food, or source of energy. By applying this type of "ecological thinking" to industry, the city of Kalundburg, Denmark has developed a unique "Eco-Industrial Model" for commercial industries—where one company's waste becomes another company's raw resource.

Waste heat from the local power plant is used to heat area homes, businesses, and fish farms. Excess steam is utilized by oil refining and biotech companies. Sulfur dioxide from the power plant smokestack is converted to gypsum, which is sold to a nearby wall-board factory.

Waste water from the refinery is used to cool and operate the power plant's smokestack scrubbers, while sludge from a waste-water treatment plant is sold to a soil remediation company that uses it to grow pollution-eating bacteria. Fly ash from the power plant is sold to nearby cement makers. All companies involved in Kalundburg's Co-Generation Project have saved money and reduced pollution, in addition to conserving water, energy, and other industrial resources.[146]

Basic Strategies for Global Waste Reduction, Treatment, and Recycling:

1) A set of environmentally appropriate global packaging standards should be created, to reduce and eliminate unnecessary packaging materials. This step alone, would result in tremendous savings of raw materials and energy, and would significantly reduce the vast mountains of packaging materials discarded into the environment every day.

2) A new set of "Environmentally Friendly" packaging technologies should be implemented. Packaging materials should be made of inexpensive materials that will degrade into biologically benign products, or can conveniently be recycled into new materials.

3) New environmentally compatible, energy efficient, strategies and technologies need to be created, which can "mine" existing landfills, separating waste materials into organic and physical waste streams—which can be further separated and refined into useful, recyclable products.

4) Creative new bioremediation technologies need to be developed, to process human and animal wastes, as well as organic, and agricultural wastes, transforming them into useful products such as organic fertilizers, fuel, building materials, and paper.

5) As soon as possible, a new global program should be funded and implemented, which employs nuclear photodeactivation technology and similar technologies to transform hazardous nuclear wastes into non-toxic waste products producing clean electrical power as a by-product of the process. All toxic nuclear waste *(including* depleted uranium), should be removed from the global environment, in order to eliminate this ominous, invisible threat to the health and well-being of innocent people—especially the children of the world!

METASTRATEGY VII

GLOBAL GUIDELINES FOR SUSTAINABLE HUMAN ECOLOGIES

**There is nothing
more difficult to take in hand,
more perilous to conduct,
or more uncertain in its success,
than to take the lead
in the introduction of
a new order of things.**

~ Niccolo Machiavelli.

The antiquated concepts of industrial-age architecture need to be transformed into entirely new and ecologically appropriate work styles and living styles for the human race. The haphazard clutter of towns, cities, and mega-metropolises which sprawl across Earth's beautiful natural landscapes, all too often represent antiquated and socially dysfunctional concepts, which have been overdeveloped to a point where they have become caricatures of their original pre-industrial village formats. Our very presence now blights—rather than complements—the beauty of nature. The crowded, noisy, and often dangerous human ecosystems we have haphazardly created are too often detrimental to the physical, mental, and spiritual health of human nature. We have trapped ourselves in unnaturally artificial surroundings, which are completely out of touch with the pulse of the natural world.

The following example graphically illustrates this phenomenon of "Human Environmental Encroachment" on the natural planetary surface. According to one account, the land area occupied by the Dallas-Fort Worth airport alone, is equivalent to a six-foot swath, which extends across America from the coast to coast.[147]

Enlightened Education and Commerce

• According to a study by the Heschong Mahone Group, students in schools that allowed in Natural Sunlight scored between 9 to 13 points higher on math and reading tests, than students from classrooms with Artificial Illumination.

• Research by the same company also discovered that Stores with Skylights averaged 40 percent higher sales than stores with solely Artificial Illumination.

~ *Earth Island Journal,* Spring, 2002, p. 18.

The Importance of Exercise, Fresh Air, and Sunlight

Although humans evolved as hunter-gatherers, in terms of the historical time-scale, modern man has only rather recently made the abrupt shift from an active outdoor lifestyle to his present sedentary and artificial indoor way of life. Physiologically, we thus exist in a state of "Natural Environmental Amnesia." Although essentially outdoor hunters and gatherers, we attempt to function normally in "Un-Natural" indoor spaces. Accordingly, we spend most of our time sitting: *sitting* in the car, *sitting* in the office, *sitting* in the bathroom, and *sitting* on the couch at home. With this in mind, the need for regular daily physical activity should be obvious.

In order to stay healthy, happy, and reasonably sane, humans require a basic minimum amount of regular physical exercise, fresh air, and natural sunlight! It is especially important to get *at least an hour* of full-spectrum sunlight *every day* to balance out the endocrine system, and stimulate the body to manufacture Vitamin D.[148] Outdoor activities such as hiking, biking, kayaking, canoeing, and swimming are far more beneficial to human health and fitness, than the repetitious mechanical exercises of a fitness-club setting. Golf and tennis offer the added advantages of combining healthy outdoor sports with wholesome social interaction. These activities all share in common the advantages of "A Healthy Interaction with the Natural Environment."

Never underestimate the importance and health value of yard work, gardening, or routine cleaning and home maintenance activities. These activities provide effective and mentally relaxing "Random Exercise," which is very helpful for releasing and balancing out the mental stress, we create during our daily workday routines.

Saving "Eco-Zombies"

• **On Earth Day, 2001, Greenwich University Ecypsychologist, Michael Cohen, announced the beginning of "Eco-zombie Rehabilitation Year."**

• **According to Cohen, "Ecozombies" are people who suffer needlessly, because they have become "Environmentally Desensitized," and thus tend to relate irresponsibly to the Natural Environment and to Life in general.**

• **The "Cure" involves simply dragging the Ecozombie outdoors to experience a "Sensory Reunion" with Nature. In Cohen's words, "Self-improvement, education and healing have always been more successful when they include contact with Nature."**

~ Earth Island Journal, Autumn, 2001, p. 3.

Indoor Pollution in Our Living and Working Spaces

A five-year study by the United States Environmental Protection Agency and Harvard University concluded that "…your home sweet home is a toxic waste hazard greater than that of the Love Canal, a chemical plant or the dump site next door." Since the average individual spends between 21 and 23 hours each day *indoors,* whatever contamination exists inside a structure has a dramatically greater effect than the pollution in the outside atmosphere.

The researchers studied indoor environments at four different locations, including three highly industrialized areas (Elizabeth, New Jersey, Bayonne, New Jersey, and Greensboro, North Carolina); and one pristine location (Devil's Lake, Idaho). Using sophisticated monitoring techniques, the air around the subjects was periodically sampled as they moved around inside the buildings where they lived and worked. Blood, urine, and breath samples were also monitored—as was the outside air. A total of 350 air samples were taken for each of the four locations.

Results of the research indicated that people living near major sources of industrial pollution had no significantly greater exposure or higher levels of body contamination, than did individuals who lived pristine environments! The study concluded that indoor levels of the contaminants under study were *at least 100 times greater than outdoor levels—regardless of the location!*[149]

"Environmental Generational Amnesia"

According to University of Washington psychology professor, Peter Kahn, children have a basic need to experience rich interactions with nature for both their physical and psychological well-being. We already know that our natural world is suffering from environmental pressures, but so too are our children, who find themselves growing up in increasingly bleak environments, far removed from the natural landscapes from which our human ancestors evolved. Says Kahn, "Even more startling is the fact that we, as adults, hardly even know this is happening."

Kahn believes that the presently impoverished state of our natural environment is at least partly due to a phenomenon which he calls, "Environmental Generational Amnesia." This concept, in essence, means that most people accept the natural environment they experience as a child as "the Norm," against which to measure the environmental degradation which they experience during their lifetime. Thus, with each successive generation, (although the amount of environmental degradation increases), the next generation accepts that degraded condition as their basic reference of what is "Natural." According to Kahn, "The upside of this is that children start fresh, unencumbered mentally by the environmental misdeeds of previous generations. But the downside is enormous, in that in that children think what they encounter is the norm in the environment. At some time you understand that the baseline is wrong, but you don't understand it at a visceral level."

The concept of Environmental Generational Amnesia emerged from Kahn's research, which examined the basic environmental values and concepts of black children in Houston, Texas. At the time, Houston was one of the most polluted cities in America. Although Kahn remembers waking up in the morning, feeling "overpowered by the smell of oil refineries," the children he interviewed in the study often said there was no pollution in Houston—even though they were experiencing the same smell that Kuhn had regarded as "overpowering."

Subsequently, Kahn conducted additional studies with a group of children and young adults in Brazil, some of whom lived in urban areas, and others in the Amazon jungle. He also included data from another group of children in Lisbon, Portugal. In all locations he discovered that similar belief patterns existed with regard to the natural environment. In other words, despite a deep ingrained need for healthy experiences and interactions with nature, people's experiences with diverse natural ecosystems are, as a whole, rapidly declining. Says Kahn, "We love nature, need nature, and are drawn to the natural world. Our connection to the natural world is so deep that some people drive for hours just to walk on the beach." In short, children require a diverse ecosystem and a variety of interactive experiences for their own well-being. Although humans have been reasonably successful in adapting to urban systems and Impoverished natural conditions, this adaptation has apparently come at a tremendous cost. Thus, for the health and well-being of present and future generations, we must, as a society, make more positive and enlightened choices with regard to preserving the natural environment in its original state.[150]

Trees Enhance San Antonio's Urban Environment

According to a recent survey by the environmental organization American Forests, existing tree cover in the city of San Antonio, Texas reduces storm water run off by 678 million cubic feet for each average storm occurrence. Construction costs to contain this volume of water runoff were calculated to be approximately 1.35 billion dollars! The study used data from satellite and aerial imaging, plus geographical information systems technology, to calculate the positive impacts of trees on the urban environment. Special software, developed by American Forests, was used to analyze the data.

In addition to reducing the amount of storm runoff, San Antonio's tree canopy also functions to remove about 17 million pounds of pollutants every year—a value estimated at 42.1 million dollars! San Antonio's urban forests store an estimated seven million tons of carbon, and sequester another 56,000 tons of carbon every year. The trees also provide another valuable function, by shading buildings from the sun. For example, air conditioners cost area residents an average of $555 a year, while residential shade trees were shown to reduce costs for an average household by $76. Based on U. S. census figures, and assuming that 68.7 percent of the city's residents have air conditioners, annual energy savings would come to about 17.7 million dollars a year!

The benefits of urban tree cover in San Antonio become even more important with the realization that the tree canopy has been steadily disappearing over the past 15 years. When Landsat images between 1985 and 2001 were compared, analysis of the data revealed a 23 percent loss in heavy tree canopy cover (50 percent or greater tree cover) for this time period. In the words of Gary Moll, Vice President of American Forest's Urban Forest Center, "The study shows that San Antonio's trees are a vital municipal asset. In the next phase of the study, we will use high resolution, multi-spectral imagery to examine these numbers more closely, and analyze them by different land cover and land use categories."[151]

Still Want to Drive an SUV?

**Switching from driving an average New Car
to a typical SUV for one year,
wastes more energy
than leaving a fridge door open for six years,
leaving a bathroom light burning for 30 years,
or leaving a Color TV on for 28 years!**

~ Worldwatch Institute New Release, March 19, 2001

The Car-Free Cities Concept

When the automobile was introduced into the major cities of
the world as the chosen means of urban mobility, it created many
unanticipated social, environmental, and aesthetic consequences.
These consequences included: 1) The disruption of pedestrian
street culture, 2) Damage to the community social fabric, 3)
Isolation of individuals, 4) Encouragement of urban sprawl, 5)
The overshadowing of urban beauty, 6) Creation of noise and air
pollution, 7) The killing and injury of thousands of people yearly,
8) The mindless waste of natural resources, and 9) The misuse of
environmental space, with a prevalence of self-serving objectives,
rather than objectives which would function for the good of the
community as a whole.

The challenge now before futurists, ecologists, and urban
planners is how to eliminate cars and trucks from our cities, while
still providing cost-effective, ecologically appropriate, and convenient transportation at a reasonable cost. One fact has become
glaringly apparent. The "Urban Automobile Ideal" can only be
supplanted if a more attractive alternative is made available! The
major questions that immediately arise are: Is the design of a Car-
free City possible, practical, and cost-effective? Would people want
to live in such a city? Would a Car-free City make practical sense
in political, social, economic, and aesthetic terms? Is it really
possible for people to free themselves from the automobile—yet

still retain the convenience, speed, and mobility offered by personal automobile transport?

To be successful, public transportation, (too often, a slow, noisy, and rather uncomfortable substitute for the car) must first be transformed into a pleasurable experience! Enough speed, comfort, and convenience must be provided to offset the expense, space, and inconvenience of auto ownership. With advanced transportation technologies it is theoretically possible to achieve this transition, so that densely populated cities can offer a truly superior lifestyle.

Venice, Italy is perhaps the best example of a Car-free City. It represents a prototypical "oasis of peace" for residents and visitors alike—despite the fact that it is one of the denser urban areas on the Planet. Through modern technology, it is possible to integrate modern transportation into the existing urban infrastructures using existing roadways and rail lines to slowly transform the key urban areas of the world into Carefree, Car-free Cities.

The design goals and standards for Carefree Cities should include the following basic elements: 1) Efficient use of existing resources, in terms of cost and available space, 2) Fast, quiet, comfortable, affordable, and efficient transportation for both people and goods, and 3) The ultimate objective of a high quality lifestyle for everyone concerned. Successful rapid mass-transit systems should thus provide access to all areas of the city, making it possible for anyone to get anywhere in a city of a million people in less than an hour. Systems should also be designed so that passengers traveling from one point to another never have to transfer more than once.

Another major feature which should be built into urban transit systems is the concept of "Intermodality." This design concept would create interconnecting terminals for passengers and goods between the different transportation modalities such as rail, bus, air, road, and sea. To enhance the quality of life, Car-free Urban Centers should also have as many public and private green spaces as possible. Such green areas would include, parks, public gardens, pedestrian and bicycle pathways, lakes, rivers, and recreational fields. The ultimate objective would to have green areas which would be accessible to anyone, within a five-minute walk from their living or working space.

Practically speaking, our love affair with the personal automobile (the "steel-encapsulated ego") must eventually be replaced by more efficient and ecologically appropriate transportation systems. Such high-tech transportation systems will eventually serve to completely remove individual auto-ownership from the equation of a high quality lifestyle—replacing it with the futuristic, non-polluting, and peaceful Car-free City concept.[152]

The Slow Cities Movement

Along with the Car-free Cities Concept, another parallel concept for urban living has quietly emerged: "The Slow Cities Movement." The Slow Cities Movement is a concerted effort by a group of 32 Italian cities and towns to adopt only those changes that will improve the quality of life for local citizens, and reinforce traditional Italian cultural values.

The Slow Cities Movement emerged in the 1980's as an offshoot of the "Slow Food Movement," a reaction to the proliferation of fast-food restaurants that had begun to invade traditional Italian towns and cities. The Slow Food Movement encourages leisurely sit-down dinners, emphasizes the value of social communication, and highlights the rich cultural, culinary and artistic local traditions of each community. Since its inception in 1989, the Slow Food Movement has continued to grow. It now boasts a membership of some 40,000 individuals, from 35 different countries!

The Slow Cities Movement also attempts to emphasize the unique cultural traditions of each community. Within this context, pedestrian areas and bike lanes have been expanded, and parks and green spaces have been improved, or added. In an effort to restore the traditional charm of each town and city car horns, alarms, and ugly signs have been prohibited or removed. Renewable energy sources, programs for recycling, and ecologically appropriate transportation systems have also been integrated into the existing infrastructures.

The Slow Cities concept is not necessarily opposed to technological progress, but embodies the simple intention of preserving the positive aspects of culture and unique traditions of each village and city. The new Slow Cities logo, proudly displayed on roadside signs, depicts a snail crawling past two buildings—one ancient, the other modern.[153]

Dubai's Incredible Man-Made Islands

Two colossal artificial islands are being constructed off the coast of Dubai in the United Arab Emirates. The Islands—built in the shape of two giant palm trees—will add nearly 75 miles new beaches to the nation's coastline.

Aware that Dubai's Petroleum Reserves represent a finite resource, Sheikh Mohammed bin Rashid Al-Maktoum, the Crown Prince of Dubai, is investing some US $3 billion to transform 80 million cubic meters of rock into a world-class tourist resort. The resort areas will contain dozens of hotels, golf courses, two marinas, a marine park with dolphins, and luxury villas which start at US $550,000. These villas will be available to wealthy foreigners on a 100-year lease basis. A curved man-made barrier reef will provide storm protection for the main residential areas, and will also serve as a marine sanctuary for visiting scuba divers and snorkelers.

Dubai's Palm Island Project is large enough to be clearly visible from space. By comparison, the only other human artifacts that can be distinguished from space are the Great Wall of China and the garbage landfill at Fresh Kills, New York.[154]

A Futuristic Floating Utopia

A revolutionary and unique Norwegian luxury liner, *GaiaShip*, is presently in its preliminary planning stages. This futuristic ocean liner tastefully integrates the concepts of visionary architecture, environmental propriety, and sociopolitical idealism. The unique design for this new cruise ship centers around a six-story glass sphere called the "Gaia Globe," which is positioned amidships. It encloses an ultra-modern conference center for world leaders, as well as sanctuaries for the world's major religions. Other amenities include a Cross-Cultural Center, and an exotic Asian Health Spa.

The futuristic concept for this "Floating Urban Utopia" was conceived by renowned Norwegian shipping pioneer, Knut Utstein Kloster, Danish naval architect Tage Wandborg, and communications expert Roar Bjerknes. The *GaiaShip* which is still in the early stages of conceptual and financial creation, is conceived as a self-financing, oceangoing center for international, political, scientific, religious, and cultural interchange. The unique new design concept has already won the endorsement of futurist, and renowned science fiction writer, Sir Arthur C. Clarke, who comments, "I think this is an excellent idea... There's nothing like a ship voyage to create a feeling of unity—as well as a better understanding of the wonderful planet on which we live." The celebrity science-fiction writer and futurist is presently assisting in fund-raising activities for the $350 million project. A Future-Studies Center on the Ship will be named in his honor.

According to the mega-project's master designers, the *GaiaShip* combines the twin aspects of a Norwegian vision—one idealistic, and one commercial. The idealistic concept is integrated into the ship's primary mission: to serve as an International Meeting Center; and also into its environmentally conscious design concept, which includes gas turbine engines, waste incineration, solar panels, and recycling systems. The objective of the GaiaShip concept is to combine leisure and self-enrichment with a global intercultural experience. If this enlightened project proves to be successful, *Gaiaship* could serve as a catalyst for revolutionizing the entire cruise ship industry.

Kloster and his partner Wandborg have already achieved a notable track record, by transforming the outmoded *S.S. France* into the *U.S.S. Norway.* By targeting an elite and erudite class of consumers, the two partners hope to insure a level of commercial sustainability, which will bring their grand seafaring visions into the realm of economic viability. Included in the commercial onboard amenities will be an IMAX Theater, a Grand Casino, a variety of World-Class Boutiques, an Internet Cafe, and a Deluxe Health Spa which offers an array of treatments from around the world. *GaiaShip* would, in essence, embody the concept of an intellectual playground, where culturally creative passengers could indulge in artistic or intellectual pursuits, and participate in philosophical and scientific studies.

Perhaps the most important concept embodied in this Millennium-scale project is the idea of *GaiaShip* as a floating "World Peace Center." Such an International Center could function effectively as a neutral venue for international peace negotiations and important meetings for key global leaders. In the words of Sir Arthur C. Clarke, "This would be an investment for a more peaceful future for us all."[155]

At the time of this Second Edition (2009), the author is working directly with the head of the GaiaShip Project, the Norwegian government, and Humanitad International to entirely re-design the original GaiaShip design into a zero-carbon footprint cruise ship, which combines computer-controlled solar-sail technology, wind power, and wave power to create a prototypical vessel that will essentially represent an entire new paradigm for marine transportation.

Intelligent Design

**With Intelligent Design
we can revitalize our cities and downtowns
using vertical zoning and optimum benefits planning.
Using Intelligent Design,
we can create pedestrian pockets in our suburbs,
where children and adults alike can grow and prosper
without the fear of a drive-by shooting,
or a daylight robbery.
With proper art and intelligence,
we can remake neighborhoods
where the kids can walk to school,
and the home office worker can take the bike
to meet the boss for lunch.**

~ Michael J. Osborne, 2001, *Lightland,* p. 44.

"Lone Star City," a Car-Free City of the Future

Visionary Architect Doug Michaels has created a circular domed megalopolis, which is designed to accommodate a population of 100,000 people. Houston resident Michels has named his futuristic urban ecosystem, "Lone Star City," after Texas—which is often referred to as the Lone Star State. The basic city design consists of a graceful circular life ring, which contains private residential sections, as well as educational, and professional areas. The city's exterior surfaces are composed of chromed steel, polished Teflon, and electronic glass. Interior vistas look out across the mile-wide interior space, over a vast central park, which contains 2100 acres of natural beauty—including forests, waterfalls, scenic waterways, lakes, pedestrian and bicycle paths, and colorful garden parks (in essence, a Natural Eden).

Beneath the life ring is a network of underground tunnels, which contains utilities, transportation facilities for the distribution of goods and services, and security and emergency centers. Deep underground is a protected deep water reservoir and a sports dome, which Michels calls the "SuperSphere." This Multipurpose entertainment venue would be home to the city's professional football team, "The Chrome."

Inspired by the rolling West Texas landscape, architect Michels sees his Lone Star City as a sensible alternative to "Suburban Sprawl." Accordingly, he has contained his self-sufficient Super-City within a massive dome, to minimize its ecological footprint, and provide a sense of community, security, and a functional integration with nature. The Super-City can either be nestled into a natural environmental setting, or incorporated into an existing urban area.

A major feature of the Lone Star City concept is the complete elimination of personal automobiles from the city's interior, thus bypassing a host of undesirable environmental factors such as air pollution, expanses of paved highways and parking spaces; and the usual costs and inconveniences associated with owning a personal vehicle, such as initial cost, depreciation, fuel, insurance, parking, and maintenance. In the words of Michels, who is the Director of the University of Houston's Future-Lab, "Maybe five years ago it would have seemed like a radical idea, but say you have an average American city of 100,000 people. The energy taken to operate that system is massive, Lone Star City will operate for at least 50 times less, so you're building a much more efficient city, and cheaper. In terms of reality, it's important to know that it's not a super expensive way to do it, but actually more economical. Its intentions are always toward reality, and improving the quality of life. Lone Star City is an idea for immediate realization."[156]

Another human-friendly feature of the Lone Star City concept is the emphasis on "Heart to Heart Contact"—a new spirit of community, cooperation and social interchange, which focuses on the city's central "Natural Commons." Although cars will not be allowed inside the Super-City, fast, efficient, and pollution-free transportation will be provided by monorails, pedestrian beltways, and compact electric vehicles. Pedestrian walkways and paved pathways for bicycles and light electric "Easy Riders" will also be part of this futuristic urban ecosystem—thus providing easy access from any one location to another. What Michels refers to as, "New Main Street" will take the form of a circular promenade, which will run around the periphery of the Central Park. Lining this promenade will be stores, coffee shops, schools, and professional service offices. Originating from the ground outside the dome will be a graceful metallic "Carapace" which arches over the central portion of the dome. This structure will contain the "Electronic Brains" of Lone Star City, the centers for city government, communications, and E-business.

The Lone Star City concept could be scaled either up or down, according to regional requirements. The domed circular megalopolis offers many advantages in its ecologically appropriate design, which is effective in concentrating a finite number of humans inside a car-free, pollution-free, and ecologically sustainable environment—while creating a minimal footprint on the planetary surface. Futuristic human ecosystems like Lone Star City could

thus be effective in eliminating unsightly urban sprawl. Such integrated and artificially intelligent cities of the future could be
designed to generate their own clean renewable energy, recycle
human and commercial wastes, and produce their own food
crops. City activities would thus take place within a healthy and
environmentally balanced system, designed for the purpose of encouraging and reinforcing creativity, cooperation, and a renewed
Sense of community spirit for all its citizens.

Ultimately, such integrated domed cities could be inter-con-
nected via fast, pollution-free mass transit systems, with rental
vehicles being available for transport to locations not yet served
by mass transit. Inhabitants of such futuristic city systems would
undoubtedly find themselves becoming increasingly transformed
by their new lifestyles, which would eliminate most of the stress
typically found in conventional cities, thus leaving them more
time for leisure activities and creative projects. By abandoning
historically redundant artifacts like the personal automobile—with
its ecologically inappropriate infrastructure of roadways, parking
lots, service stations, auto manufacturing plants, sales lots, and
garages—more time and space would become available for creativity, and "the Pursuit of Happiness" for all citizens young and
old alike. In the spirit and joy of his new and creative social adventure Doug Michels declares, "Live beyond the memory, create
the future!"[157]

Jacque Fresco's Brilliant Vision for the Future of the Human Race

Visionary Engineer, Inventor, Architect, Philosopher, and Futurist, Jacque Fresco, has devoted most of his life energies to creating a conceptual future world, where the resources of Planet Earth
would eventually become the common heritage of every member
of the human race. He and his professional associate, Roxanne
Meadows, envision a new, enlightened version of global civilization, where science and technology would seamlessly integrate
human and environmental concerns, providing social justice and

equal opportunities for everyone. Jacque Fresco's brilliant futuristic paradigms provide a set of future scenarios, where global society transcends the endless cycles of economic boom or bust, and eliminates the specters of disease, poverty, hunger, as well as the bloody conflicts that have retarded the progress of human evolution.

Fresco is convinced that the major problems facing society today are mainly of our own creation. The reflective ghosts of our own past actions, which have returned to haunt us with a vengeance! His basic thesis rests on applying the knowledge and technologies we already possess to enhance the quality of human life on earth, while recognizing the necessity for achieving symbiotic interrelationships with all living systems, and the critical importance of protecting and enhancing the natural environment. In order to accomplish these objectives, he suggests we must work to intelligently re-arrange human affairs until they are "in compliance" with Earth's natural systems and resources. In the introduction to his most recent book, *The Best That Money Can't Buy: Beyond Politics, Poverty, and War,* Fresco states, "We call for a straightforward redesign of our culture, in which the age-old inadequacies of war, poverty, hunger, debt, and unnecessary human suffering are viewed not only as avoidable, but also as totally unacceptable. Anything less simply results in a continuation of the same catalog of problems inherent in the present system."[158]

Jacque Fresco's brilliant series of books and videos make full use of architectural modeling and visual media technologies to demonstrate how artificial intelligence, natural resource conservation, modular construction systems, clean energy, efficient mass-transit systems, advanced food-production technologies, futuristic waste-processing systems, and nanotechnology can all be integrated into a new techno-social architecture, which operates synergistically with the laws of nature and in harmony with the global ecosystems. His enlightened vision of "Globalization" serves to empower every person on the planet, providing them with the opportunity to rise above traditional barriers of oppression, and to reach their full potential—physically, intellectually and spiritually.

In his book, *The Venus Project: The Redesign of a Culture,* Jacque Fresco presents a world of virtually limitless possibilities and an adventurous new social direction for the human race. In the futuristic scenario he has created, the current value of individual purchasing power (based on a trade-off of human effort for monetary wages) would become a historical anachronism, since human labor would be replaced by artificially intelligent machines. This, In light of the fact that people everywhere are losing confidence in the ability of their political leaders to provide workable solutions for their social, economic, and environmental concerns. The Venus Project proposes a fresh new vision for global society—a vision dedicated to both human and environmental concerns. Fresco envisions a future where his advanced technologies are established in ways that are both practical and feasible, so that they can be applied to benefit everyone.

Fresco feels that society as a whole has so far been unable to make the necessary adjustments to properly master the technologies they have developed. Too often they have used these technologies for purposes of mass destruction. This failure of the human integrity, which he refers to as "Neural Lag," has caused the general public consciousness to remain enmeshed in antiquated thought patterns, which remain dominated by the concepts of "Conquest," "Scarcity," and "Greed." He points out, that if the same energies that went into the Manhattan Project could have been re-channeled into improving human society on a global scale, and in anticipating the needs for the future, we would presently exist within a much more symbiotic relationship with other nations, cultures, and the natural world.

Jacque Fresco also highlights the fact that the outmoded political and economic systems that presently dominate the contemporary global scene are woefully inadequate for applying the positive benefits of technology in ways that will achieve the greatest good for the greatest number of people. He envisions a new form of "Participatory Democracy," where each individual would contribute according to his or her unique talents and level of competency. Such a major shift in consciousness could be accomplished only within the kind of positive learning environment that would eventually enable everyone on Earth to access the educational and cultural amenities that only a prosperous, enlightened, and innovative society can provide. Through computers, telecommunications, and virtual reality systems, information resources would be shifted from their historical position of being concentrated within the universities, churches, and libraries, toward a state of decentralization—where vast digital information resources and virtual reality experiences would be within the reach of everyone, even in the remotest corners of the Earth.[159]

Fresco proposes a major paradigm shift, which would begin to move the human social consciousness away from its present capitalistic theme, which promotes "Survival of the Fittest," the "Conquest of Nature," and encourages the "Plundering of Earth's Natural Resources" without regard for ecological justice, or the consequences of such actions for future generations. He also feels that our present monetary system evolved over time, as a device for controlling human behavior. Instead, he proposes we work to establish a "Resource-Based Economy," which would incorporate the free exchange of clean, environmentally appropriate technologies, and simplify commerce and technology by adopting global standards for all aspects of life. This New Economy would also encourage creative and intellectual exchange between all cultures and religions. He goes on to state that all social systems, regardless of their political, religious or social beliefs, ultimately depend on natural resources (i.e. clean air and water, fertile soil, appropriate technologies, and human brain power) to maintain a reasonable and comfortable standard of living. In Jacque Fresco's words, "The real wealth of any nation lies in its developed and potential resources, and the people who are working toward the elimination of scarcity, and toward the creation of a more humane life-style. This can be accomplished through the intelligent applications of Science and Technology."[160] In its quintessential form, a resource-based economy uses existing resources instead of money and establishes an effective system for distributing these resources in the most efficient and equitable manner possible, and for the entire population of Earth.

In Jacque Fresco's vision for the future, computers and intelligent machines would be the primary "Agents of Change." These AI's would employ sophisticated sensor arrays to monitor and manage resources. Complex Cybernetic Systems and Intelligent Machines would monitor environmental parameters continuously, and on a global scale, providing protection from the unauthorized "pirating" of natural resources, especially in protected forests, coral reef preserves, and endangered marine ecosystems.

Abundant, non-polluting, cheap energy could be provided by upgrading our existing solar, geothermal, and wind power technologies. Other new technologies could be developed to tap the vast resources of tidal and ocean currents. The oceans of the world could also yield a nearly inexhaustible wealth of fresh water and mineral resources, which could be "mined" directly from seawater.[161]

A particularly fascinating aspect of Fresco's Futuristic Paradigm for human society involves the creation of "Integrated Cities," which apply new and innovative applications of architectural design to build cities which would constitute synergistically functional units. These "Intelligent Cities" would be tastefully blended into the natural environment. The city ecosystems are circular or star-shaped in design. A central domed area would enclose shopping areas, theaters, sports arenas, entertainment venues, health facilities, and commercial centers, as well as centers for communications, education and childcare. The central dome would also contain facilities for control and distribution of electrical power, transportation, and communications services. Outside the central dome, a ring of smaller domes would house art, music, science, research, exhibition, and entertainment centers. The next outer perimeter would contain a circular ring of residential towers, set like jewels into a complex of parks, lakes, streams, forest preserves, and recreational facilities. Outer ring sections would contain areas for power generation, water storage, food production, waste processing, and sewage treatment. The transportation network would consist of a system of horizontal, vertical, radial, and circular conveyors that would move passengers and goods efficiently and safely from one area of the city to another. This integrated transportation system would completely eliminate automobiles, roadways, garages, and parking areas from the city complex.[162]

Fresco applies his creative architectural concepts to conserve energy, and to create minimal ecological footprints for all his cities and buildings. To this end he envisions creating Colossal Intelligent Machines which could drill tunnels through the earth for transportation, and dig canals for purposes of irrigation and water transportation. Similar giant machines would assemble huge residential and industrial complexes from mass-produced modular components. Other "Intelligent Structures" are designed to be self-erecting. Fresco's futuristic skyscrapers, made from pre-stressed concrete, reinforced with carbon fibers, would take the form of huge tripod-like structures, consisting of three massive elongated tapered columns, 100 feet wide and nearly a mile high. These artistic residential towers would eliminate unsightly and inefficient urban sprawl, thus insuring that land would be available for parks, wilderness-recreation areas, and intensive food-production facilities.[163]

Jacque Fresco's Vision for the Future foresees a time when the patriotism and national pride (which presently obscures the contributions of less dominant nations) would no longer be relevant to the new, emerging culture. His educational system for the future employs "Cooperation" and "Hands-On Experience" to develop confidence and improve social skills. Learning centers would stress the importance of communications skills and human values. This new educational system would be free from the restrictions of the moribund educational institutions of the past, as well as from indoctrination by special interests. According to Fresco, "Children would understand that each individual can take an idea only so far—each contribution motivates and encourages others. Ideas grow and expand like crystals into varied and complex patterns. With this better realization of our interdependence with one another, self-centeredness gradually disappears." Education would thus become a "Continuous Lifelong Adventure," with residents of globally networked communities, being educated from birth to become "Global Citizens," without sacrificing either their individuality or their personal freedom. Fresco encourages us all to envision a future where thousands of individuals can live and prosper—while thinking and creating to their fullest potential. In such a Future World, the majority of humans would actively participate in improving the conditions on Earth, instead of simply toiling to make a living.[164]

Basic Strategies for Developing Sustainable Human Ecologies:

1) Functionally integrated "City Ecosystems" should be created to replace dehumanizing urban sprawl with personalized living spaces, configured in the form of compact communities and hi-rise clusters. Within these car-free, people-friendly human ecosystems, tree-lined pathways could be created for pedestrians, bicycles, and personal electric vehicles.

2) Unique new urban ecologies would be designed with clean renewable energy, quiet mass-transport, and efficient waste management systems. Parks, artistic green spaces, and "Green Commons," would be created to complement personal living space, thus encouraging outdoor activities and a new sense of community spirit.

3) Comfortable, quiet, and efficient mass-transit systems would provide links between other cities and community centers. Rental vehicles would be available for travel to areas not served by public transit, so personal vehicle ownership would not be necessary.

4) A multi-function domed sports complex would provide venues for stadium events. Lakes, rivers, and forest parks outside domed city areas would provide space for outdoor recreation. Other nearby areas would be designated for water storage, food production, and waste and sewage processing.

5) Futuristic city ecosystems like Doug Michels' "Lone Star City" and Jacques Fresco's "Venus Project," have already established new, environmentally appropriate, futuristic paradigms for human habitation. These "Urban Ecosystems of the Future" would tastefully combine living and working space with natural space, to create a "Re-Naturalized" sense of community spirit for their citizens.

METASTRATEGY VIII

CLEAN, ENERGY-EFFICIENT GLOBAL TRANSPORTATION SYSTEMS

**The Transportation Network
is to the Economy as a whole
as the arteries and veins of the circulatory system
are to the body.
It ought to be capable of moving goods and people
where they are needed,
in the most timely and effective way.**

~ Lyndon LaRouche's Committee for a New Bretton Woods, p. 19.

The Hypercar Concept

The Hypercar incorporates advanced concepts of ultra-light construction, a low-drag design, a hybrid-electric engine, and efficient new technologies to achieve three- to five-fold increases in fuel economy, performance, safety, and affordability, when compared with conventional vehicles.

The Hypercar concept was developed by researchers at the Rocky Mountain Institute. Initially, a hybrid-electric drive train will be used in conjunction with an advanced internal combustion engine. To reach its full potential and eliminate polluting exhaust entirely, however, the Hypercar will be powered by hydrogen fuel cells.

A spin-off company, Hypercar, Inc., is presently developing a design and manufacturing process that will revolutionize automotive structure. New features will include durable, rust-proof bodies, which are over 50 percent lighter, have twice the stiffness of conventional steel, and improved crash-protection capabilities. Unique state-of-the-art electronics will improve performance, safety, and reliability—while reducing mass, mechanical complexity, and basic cost.

221

Hypercar Inc's product development team is global, and includes some of the best minds in the industry. Their innovative product development approach was adopted from aerospace methodology, originally conceived at the Lockheed Martin Skunk Works.[165]

Driving to the Sun and Back

**Every day, in Texas,
Texans drive over 180 million miles,
or to the Sun and back,
simply to accomplish
their daily tasks and recreational activities.**

~ Michael J. Osborne, 2001, *Lightland,* p. 44.

A Compressed-Air Powered Car

In October of 2002, a unique new concept vehicle was launched in South Africa; a vehicle which offers exciting possibilities for clean and efficient, urban transportation. The new car, created by the Johannesburg Company, e.Volution, has been designed to run on compressed air. The e.Volution prototype made its global debut at the Auto Africa Expo 2000. The vehicle was presented as the first viable alternative to cars that run on fossil fuels.

For years, researchers have been working to produce ecologically "friendly" vehicles. The compact, but sporty, air-powered car will sell for about US $10,000. The body of the new vehicle weighs only 1,540 lbs. (700kg), with the engine weighing in at a mere 77 lbs (35 kg). The Car is designed to be driven for up to 10 hours in an urban environment—at an average speed of 48mph (80kph).

The creators of the new e.Volution claim it will be possible to merely plug the vehicle into any standard electrical outlet to "fill it up." The downside, however, is that this process could take up to four hours. The manufacturers envision that, with commercial air stations this recharge would only take three minutes. Alternatively, businesses and fleet owners could operate their own air stations, with the advantages of completely eliminating the fire hazards, environmental pollution, and regulations associated with the storage of gasoline and diesel fuels.

The first models of the e.Volution compressed-air car were expected to be on the streets by late 2000. Two factories have already been built in France, with five factories being planned for Mexico and Spain, and three for Australia. After France, South Africa is slated to be the second country to open a factory. They expect to begin producing the new cars by 2002. Another company, Zero Pollution Motors, plans to produce compressed air cars in the United States (www.zeropollutionmotors.us).

A major objective of the project is to cut costs for the compressed-air car by serving consumer markets directly, thus avoiding major shipping expenses. This arrangement would thus create jobs within the region where the vehicles will be sold. From an environmental perspective, if the new non-polluting motor design catches on and the technology continues to be refined, air could be compressed and stored, using clean alternative energy sources. Compressed-air-motor technology could also easily be adapted for other non-polluting applications, such as marine engines, powered bicycles, scooters, lawn mowers, and portable tools.[166]

Author's note: Although the compressed air concept has certain merits in terms of clean power, there are problems associated with the energy source used to charge the vehicles (i.e. Is it non-polluting and sustainable?), as well as other problems with limited range of these vehicles per charge, and the costs of establishing and maintaining a network of charging stations.

StarPort: An Airport Design for the Third Millennium

A typical major metropolitan airport consumes up to 500 million gallons of fuel a year—nearly half as much as the vehicle traffic in a large city. Big-city airports also typically create over 50 percent of the air pollution in large metropolitan areas. For example, a Boeing 747 uses over 500 gallons of fuel, just to taxi from landing to the gate (enough fuel to power a car for a year!). For 1,000 taxi-to-takeoffs, 12 million gallons of fuel are consumed, enough to power 200,000 cars for a day. Jet engines are notorious for their poor fuel efficiency on the ground, as *only four percent of the* fuel burned actually moves the aircraft. The rest is spewed into the air as polluting exhaust gases!

An innovative new airport design, called "StarPort," has the potential to save up to 300 million gallons of fuel per airport per year. The new design occupies only about a third the space of a conventional airport, yet yields about four times the revenue. Since StarPort design incorporates inclined runways, running parallel to each other, gravity slows landing aircraft and speeds up takeoffs. Runways are designed to be wider at touchdown points and are slightly concave, thus helping aircraft to stay centered.

A two percent incline for the 6,000-foot runways raises the runway midpoints to a height of 120 feet above the surrounding landscape. This allows the central terminal to be as high as a ten-story building, so that several floors of ticket counters, restaurants, hotels, meeting spaces, and parking lots can easily be incorporated. Parallel runways allow for simultaneous takeoffs and landings, with the inclined runway design reducing taxiing distances by as much as 80 percent. If small electric motors were installed on aircraft to pre-rotate wheels prior to landing, this would eliminate wheel burn and reduce touchdown shock. This, combined with the inclined runways, would eliminate the need for noisy 30-second breaking thrusts, which typically burn 300 to 500 gallons of fuel for every landing.

Airport customers normally endure an average 1.5-mile walk from parking lots to departure gates. With the compact StarPort Design, passengers could travel directly from underground parking lots to departure gates *in less than six minutes!* For an airport the size of Denver, this would reduce the average curb-to-counter passenger travel by 80 percent. Instead of the usual vehicle congestion associated with conventional airports, StarPort would allow traffic to enter and leave the airport in four directions.

The compact design of StarPort would allow it to be built on smaller existing urban airport sites—abandoned previously, when longer runways were required for heavier aircraft. If the 2,000 new airports which are currently on the drawing boards were of StarPort design, fuel savings would amount to approximately two billion gallons of fuel per day—more than 1,000 times the amount that could be extracted from the Arctic Wildlife Refuge![167]

Sky Cat Airships

Sky Cat is a unique hybrid airship, which combines two existing transportation technologies (one lighter-than-air and one heavier-than-air principle) in a unique new way. In this new concept, a semi-rigid aerodynamic lifting body is combined with a stable twin-hulled catamaran design. This advanced design concept combines the versatility of a Hovercraft, with high-tech materials and light weight propulsion technologies. Near the ground, Sky Cat airships function in a Hovercraft Mode—allowing them to set down either on land, or water. A unique bow thruster system enhances maneuverability, so the airship can land nearly anywhere, with minimal crew and mooring facilities.

Sky Cat 20 (the smallest model), is 266 feet long, and offers a spacious cabin with wide seats, and three- by six-foot picture windows. It can carry up to 82 passengers, at a cruising speed of 75 knots, and altitude of up to 9,000 feet. It has a range of 2,000 nautical miles. Larger Sky Cat models range up to 1,000 feet in length, with cargo capacities of up to 10,000 tons, cruising speeds of up to 100 knots, and ranges of up to 2,000 nautical miles. These larger aircraft are designed to carry cargo over moderate to great distances, at rates significantly lower than for conventional aircraft, and only slightly higher than for ocean freight. Since conventional airport facilities are not required, cargo can be picked up and dropped off nearly anywhere, opening endless possibilities for economical and energy efficient direct product-to-customer delivery.[168]

An American Maglev Network?

A Network of Maglevs could shuttle passengers between American cities at over 300 miles per hour, using far less energy and time than automobile and air travel. One could go from a downtown office to a weekend cottage 100 miles away in just 20 minutes, or take a portal-to-portal trip of about an hour and a half between the air-shuttle cities of Los Angeles and San Francisco, Washington and Boston, or Chicago and Minneapolis.

~ John L. Peterson, 1994—*The Road to 2015: Profiles of the Future*, p. 169.

"Aerobus" A Futuristic Mass-Transit System

Aerobus is an aerial, self-propelled, modular light-rail transport system, designed to run on aluminum tracks, which are suspended from steel cables. It is silent, safe, climate-controlled, and environmentally friendly. It can run at speeds of from 60 to 100 kilometers per hour, and can be installed above existing freeways, roads, or rail lines.

Narrow steel or concrete pylons support the cables, which can route Aerobus over freeways, buildings, rivers, and other obstructions, that would prohibit the use of more conventional transportation technologies. Thin, unobtrusive pylons, normally spaced 200-800 meters apart, can be placed up to 600 meters apart; a span 10 to 15 times greater than for the relatively massive monorail columns.

Modular units may be added or removed according to passenger traffic, with up to twelve modules being combined to make a "Train," which can carry 400 people at 80 kilometers per hour. By spacing trains at one-minute intervals, carrying capacity equals 20,000 passengers per hour, per direction.

Aerobus modules can be conveniently adapted for freight, by using unique bottom-loading containers which facilitate loading and unloading. Minimal disruption of highway traffic is required for pylon installation, and the problems normally associated with railway traffic crossings are completely avoided. The basic cost for the Aerobus is significantly less than for traditional monorails, and Aerobus systems can be built much more quickly.

The Aerobus concept was initially developed 30 years ago by Swiss inventor and engineer, Gerhard Mueller. Early commercial prototypes carried over 2.5 million passengers. The latest version of Aerobus incorporates four generations of technological refinements, and has already received favorable approval from the United States Department of Transportation, and the Urban Mass-Transit Association. The First of these modem Aerobus systems will soon be operational in Chongqing, China.[169]

Author's note to the second edition: Although in the final analysis this project failed to materialize, the fascinating story written in Forbes Magazine can be read at the following link: (http:www.subways.net/china/chongqing.htm.).

Table 3 — Cost per Mile of Mass Transit vs. Aerobus

Mass-Transit	Cost per Mile
Heavy Rail	$75 Million
Light Rail	$45 Million
Monorail	$30 Million
Aerobus	$10-15 Million

Cruise Ships: Polluters of the High Seas

Many unregulated, foreign-flagged cruise ships, which ply the global shipping lanes have, for years, been responsible for serious pollution violations. As they crisscross the planet's oceans, they spew forth dense black clouds of toxic atmospheric pollution—the result of burning dirty bunker fuel. Cruise ships continue routinely discharging oil, plastics, paint, hazardous chemicals, garbage, and sewage into the world's oceans. With their garbage, chemical spills, sewage, and smog, it has been said that these "Floating Cities" often have more in common with third-world factories, than with the "Love Boats," depicted on TV and in glossy travel brochures.

The Federal Standard for coliform bacteria is 200 colonies per 100 ml of sampled wastewater. Samples from the ships' sewage treatment plants, however, often register in the millions! Although every ship has a "certified" sewage plant, *not one* of 22 ships tested in 2001 passed the test when their effluent was checked! "Gray water" from baths, laundries, and galleys was found to be as contaminated with fecal waste as the "black water" systems. One Alaska regulatory official described these ships as "Floating Cesspools."

Hopefully, this deplorable situation will soon begin to improve. A new bill, signed into law on December, 2000 requires the Coast Guard to develop workable monitoring and reporting programs for all cruise ship discharges. The main objective of this program is to ensure that the cruise ship industry is held accountable to the same rules of disclosure and protection that govern other discharging industries. In order to protect our "Global Commons," similar programs need to be established to monitor pollution and enforce environmental regulations for cruise ships, and for freighters, tankers, and other seagoing vessels.[170]

A Futuristic High-Speed Rail Commuting Scenario

Consider the possibility of commuting the 450 miles from Boston to Washington, DC in 90 minutes, at a price cheaper than today's Metroliner! For the one million people who make this journey every year, this Vision could become Reality—through the development of high-speed Maglev Mass-Transit Systems. In this Futuristic Scenario, passengers travel from their homes, via a network of 50-mph Maglev commuter lines to a central station, for departure on the 7:30 AM express line to Washington. Once aboard, Passengers relax in quiet comfort—free from the hassles of rush-hour traffic and finding parking spaces. They can enjoy breakfast, read the morning paper, make phone calls, or check e-mail via high-speed internet connections. From the Central Station in Boston, the train accelerates to a cruising speed of 300 mph. After a brief stop in New York City at about 8:15, the train continues, arriving in Washington DC at 9 AM—just in time to begin of the working day.

~ Lyndon LaRouche, 1998—*The LaRouche*
Program to Save the Nation, p. 26.

Fast, Efficient, and Environmentally
Compatible Mass-Transit Railways

From an economic perspective, marine shipping is the cheapest mode for transporting commodities. In moving freight over long distances, (especially between continents and island nations), shipping is presently limited by speed. In the absence of navigable rivers or canals, marine shipping is unsuited for efficient transcontinental transport of goods and passengers. Thus, for most purposes, railways offer the most practical and cost-effective shipping method.

The usual way to compare basic transportation costs is to calculate the weight and speed which each transport mode can move people or goods. Using this evaluation method, a double-track rail line with only three trains per hour at 60 mph, can move the same quantity of goods as far in one hour, as a fleet of 330 20-ton trucks. Labor and energy costs for the two modalities are about equal. Ecologically speaking, however, the double-track rail line requires only about *one-twelfth the land area* of a conventional interstate highway. Since truck speed is limited by the technology of the internal combustion engine, the practicality of rail transport increases dramatically with speed—the upper limit for conventional diesel locomotives being about 125 mph.

High-speed passenger rail systems have already been developed to function effectively at speeds in excess of 200 mph. These high-speed systems, developed over the past two decades, include France's *a Grande Vitesse,* and Japan's bullet train *(Shinkansen).* The next generation of high-speed rail transport, now being pioneered in Germany, France, Japan, and China, uses different variations of magnetic levitation technology, where the trains glide along on a cushion of air between the undercarriage and the track. These new high-speed rail systems can carry either passengers, or freight, or any combination of the two. Trains run on time, and are independent of weather conditions, and aircraft runway congestion. Train stations also create far smaller environmental footprints than do airports. Since trains can pick up and deliver passengers or freight directly at strategic urban locations, time-consuming airport-to-city transfers are eliminated.

Outside the U. S., high-speed Magnetic Levitation Systems (MLV's) are rapidly becoming practical reality. In Germany, for example, the Basic System has already received governmental approval. The initial phase of this project will connect airports in Cologne-Bonn, Dusseldorf, and eventually Essen. The European program is about seven years ahead of the Japanese program, where a new MLV system is scheduled to begin passenger service in 2010, running in the densely populated corridor between Tokyo, Nagoya, and Osaka. Germany and Japan have also concurrently developed lower-speed systems, like the German *M-Bahn;* a 50 mph MLV urban transport system, which operates in Berlin. HSST Corporation, the manufacturer of these rail vehicles, has provided vehicles for demonstration systems in Tsukba and Yokohama, Japan, and in Vancouver, Canada. Thus, practical maglev technology has already been developed for different transportation functions—ranging from short-distance, high-speed commutes at 60 to 250 mph, to intercity journeys at speeds in excess of 310 mph.

Germany's *Transrapid TR-O7* can carry up to 200 passengers at speeds of up to 310 mph. With units spaced at one-minute intervals, this rapid rail system can transport between 10,000 and 20,000 passengers per hour. Japan's commercial maglev train, the *MLV002,* will have 14 cars, capable of transporting 900 passengers per trip. It is intended to move from 75,000 and 100,000 persons a day between the urban centers of Tokyo and Osaka. Both Germany's TR-07 and Japan's MLU 002 will incorporate so-called "Passive Systems," however the German and Japanese Systems employ different electromagnetic principles for the suspension, propulsion, and guidance of their respective vehicles.

Maglevs typically use two different types of propulsion and guidance systems. The first type has propulsion and guidance systems which are propelled and controlled from the track (Japan's HSST Model). The second type (Germany's TR-07 and Japan's MLU 002) uses "Passive Systems," where propulsion and control systems are located onboard the vehicles. The German Transrapid is based on an electromagnetic suspension system, in which the vehicle's undercarriage wraps around the guideway, pushing the vehicle up and off the track. The Japanese system uses repulsive forces to lift the vehicle off the guideway. This latter system uses a wheeled undercarriage for "liftoffs" and "landings," since passenger cars can only levitate at speeds above 25 mph.

Where does the United States stand in this futuristic high-speed rail transportation scenario? Paradoxically, high-speed maglev trains are essentially modernized versions of a basic technology developed in the USA back in 1971, when extensive research was carried on by Stanford Research Institute, the Ford Motor Company, Rohr Industries, and the Boeing Corporation. In 1974, a world speed record was set by a prototype linear induction motor vehicle at a Colorado test facility. In 1975, however, federal funding for this developmental program was terminated. At that time, the United States was *at least a decade ahead of the rest of the World* in high-speed electric rail development. Now—it is completely out of the picture!

When compared with air travel over distances between 90 and 200 miles, high-speed maglev systems can carry *twice as many passengers* at *half the cost* of aircraft such as the Boeing 737. In addition, rail systems are safe, and offer greater passenger comfort, enroute mobility, and convenience than does air travel. Rail systems are also more ecologically appropriate in terms of air pollution and overall efficiency. Unlike air transport, rail systems can move passengers or freight *directly* to and from urban centers and other destinations, without the hassles of traffic congestion between airports and city centers. Costs for maglev systems are projected to be *cheaper* than for conventional rail systems, with operation and maintenance costs being projected at 5.2¢ per passenger mile (in 1988 dollars). This is considerably less than passenger-mile costs for today's metroliner, which runs between 16.2¢ and 36¢ per mile—depending on the type of accounting analysis.

In terms of passenger hours saved (i.e. commuting to and from airports, waiting, and taxiing time), *maglev systems would actually pay for themselves.* It has been estimated that approximately 40 billion dollars a year is lost to traffic delays in America's eight most congested urban areas alone. This money could put to much better use by financing the construction of 3,000 miles of maglev rail networks a year. The jobs and industrial stimulation created by the development of a high-tech national rail system would return far more in terms of increased productivity, than construction would ever cost.[171]

Basic Strategies for Clean, Energy Efficient Global Transportation Systems:

1) Energy inefficient, space-wasting personal vehicles should be replaced by mass-transit systems which are cost-effective, space and energy efficient, non-polluting, quiet, and comfortable.

2) Internal combustion, fossil fuel engines should be replaced by efficient, non-polluting hybrid, and alternative Energy power sources.

3) More efficient use of existing highway space could be achieved by eliminating private, single-passenger vehicles, in favor of space- and energy-efficient mass transit vehicles, which would use existing highway rights-of-way—thus effectively creating usable highway space for commercial trucks and automobiles.

4) By adding elevated light-rail mass-transit systems to existing highways systems, park-and-ride passengers would have convenient access to quiet, comfortable, and energy efficient mass transport for commuting between home and commercial centers.

5) Worldwide Intermodality standards should be created, to facilitate the efficient transfer of passengers and freight from one transport mode to another. Multipurpose "Hub Terminals" would thus be designed to interconnect air, rail, sea, and land transportation networks.

6) Entirely new and energy efficient transportation technologies need to be brought into commercial reality. This would provide safe, reliable alternatives for air travel (an expensive, polluting, and ecologically inappropriate alternative to the high-speed rail networks that already exist in many European countries). Hi-tech airships do not require expensive airport complexes. Long-range travel could be upgraded with massive public funding programs to develop high-speed intercontinental rail systems, advanced marine propulsion technologies, transcontinental air ships, and hypersonic space vehicles, which would use futuristic, non-polluting technologies.

METASTRATEGY IX

A GLOBAL SHIFT FROM
CONSUMPTION-DRIVEN
TO ECOLOGICALLY-APPROPRIATE
LIFESTYLES

The Earth is under siege by us.
We have been transformed
into Insatiable Consumers.
The issue before the Eskimos
is not whether they need Refrigerators,
but how they can buy them
and use them.
For most humans,
Consumption
is the central purpose of Life.

~ Carlos Hernandez, R. Mayur, 1999, *Pedagogy of the Earth.*

Our Over-Consuming Culture

Since the beginnings of recorded history, humans have been using up Earth's finite resources far faster than natural processes can regenerate them. We consume genetic resources 1,000 to 10,000 times as rapidly as natural evolution creates them. Scientists surveyed at New York's Museum of Natural History have indicated that we have entered the fastest mass extinction period in the history of Earth—even faster than when the dinosaurs disappeared![172]

We consume topsoil much faster than nature can restore it; with some 30 billion tons of this precious resource being lost each year. Loss of topsoil has a similar effect on Earth's Biosphere, as a person losing blood—only so much can be lost, before the body dies.[173]

235

Affluenza: The New *Affliction du Jour!*

• **If you work longer to pay for things you have less time to enjoy, you may be suffering from "Affluenza."**

• **The problem is, the more you have, the more you tend to want, which creates an Addiction that has been referred to as "Viral Overconsumption."**

• **Despite the fact that Personal Consumption Expenditures have tripled over the last 50 years, the percentage of people claiming to be "very happy" has declined by almost 50 percent since 1973.**

~ John De Graaf, D. Wann, T. H. Naylor, 2001, *Affluenza: The All-Consuming Epidemic.*

Exotic Chinese Dining: The Irresistible Lure of Eating Endangered Wildlife!!

Located just outside of Lianbian, a small village on the outskirts of Guangzhou, is a trendy restaurant with the nonsensical name, "Youth No. 1 Village Wild Flavors Restaurant," where the newest fad is feasting on weird and expensive wild animals.

The hundreds of wild animals confined in cages, stacked high along the walls include: herons, cranes, baby deer, lynx, bobcat, squirrels, snakes, and scorpions. These animals are all destined to end up fried, stewed, or dunked into hot pots, as part of the day's "Businessman's Lunch." The restaurant's theme is "Chairman Mao's Cultural Revolution." Its walls are decorated with huge red banners, bearing messages like, "Long live the proletarian revolutions." Waitresses are dressed in uniforms of the Red Guard. Owner, Liu Zhenhua, sporting a shag haircut, two ear rings, and a trendy cellphone says, "Now we are prosperous and can eat anything we want, so we can remember past suffering, while enjoying our current good fortune."

Wild animal restaurants are just one of the many current Chinese consumer crazes which range from skateboarding, to fine foods and wine—trends that have swept China recently, as rising incomes have resulted in excess time and money for ambitious young businessmen to spend. As a result, the road leading north out the provincial capitol, Guangzhou, is cluttered with gaudy restaurants, whose signs would seem better suited for zoos, than for trendy dining spots.

"Wildlife restaurants are a sign of our prosperity," says Kuang Zuoqiao, a 51-year-old farmer, who has come with two of his relatives to Liu's establishment to enjoy cheap cigarettes and expensive chunks of wild boar meat. "It's fun and exciting to see what new animals taste like." He claims he and his friends splurge on wild animal meals two to three times a month, stating, "When you see an animal, it's only natural to wonder what kind of flavor it has."

At the restaurant's front entrance, the well-dressed customers view the cages stacked along the walls—picking out their favorite lunches. The featured house specialty is "Dragon, Tiger, Phoenix," a stew made from snake, wildcat, and crane; popular because of its supposed traditional health benefits, as well as its flavor. According to one of the waitresses, rats are also popular—but only during the winter months, since they carry too many diseases in the summer. However, customers like Kuang Zuoqiao seem to feel that, like fine wine, a well-prepared rat knows no season! Says Zuoqiao, "To me, it doesn't matter what time of the year."

There *are* those in China who feel that the people of Guangdong have carried the business of "Extreme Eating" too far. Accordingly, the Central Government, responding to the fact that wealthy locals are paying exorbitant prices to dine on protected and endangered species, has begun a campaign to crack down on this unnecessary abuse against nature. Just recently, the owner of a local restaurant was sentenced to five years in prison for serving Pangolin, an endangered species of anteater, which is a prized black-market item in Asia—both for its meat, and for its unusual scales which are sometimes used for guitar picks. Restaurant owner, Liu claims he is careful to obey government laws, and often buys animals from Southeast Asia that are not available in China. According to one waiter at the restaurant, "Rare owls and crocodiles used to be popular, but you can't get them any more, because the government has banned that."[174]

This trend is just one example of environmentally destructive consumerism. The list of consumer fads, which have decimated wildlife over the centuries, includes: ambergris from whales for use in the fragrance industry; ivory from elephants, walruses, and narwhales; beaver hats; ostrich plumes and boas; crocodile bags and shoes; snake belts and hatbands; turtle meat and shell crafts; and the pelts of foxes, leopards, lynx, zebras, and lions. The Asian quest for exotic game foods, traditional medicinal products, and aphrodisiacs from bears, tigers, rhino horns, shark fins, and sea horses, have pushed many of these species to the brink of extinction, distorting the balance of the major ecosystems where they once thrived.

The hope for future generations lies in educating young people and the general public everywhere, so that each and every citizen of the global community will learn to respect the natural environment. In doing so, they will experience the profound kinship and intrinsic rewards that these wild species return to the human consciousness, just by being there!

Viagra Saves the Rhinos

Viagra has done more to save endangered Rhinos, than the World Wildlife Fund! Since Viagra gets it up better than traditional oriental remedies (including Powdered Rhino horn). Asians have been quick to make the switch, so poachers have lost a major market share.

Now...if we could find an alternative for Yemeni Ceremonial Dagger Handles, we could eliminate the remaining Major Cause of African rhino deaths.

~ *Whole Earth,* winter, 2001, p. 106.

Seahorse Aquaculture

New Zealand researchers have recently developed effective new techniques for culturing seahorses in enclosed aquaculture systems. This new program will hopefully help alleviate the severe fishing pressures on wild stocks, and begin to restore the biological balance in the natural environment. During recent years, sea horse populations in areas like the Philippines have declined drastically. Seahorse populations everywhere in the world are under intense fishing pressures, due to the high prices in Asian pharmaceutical markets, and the demand for aquarium specimens and dried souvenirs.

In addition to their bizarre appearance, seahorses are unique in that they demonstrate a phenomenon called "Parental Role Reversal." After a stylized mating courtship, the female seahorse transfers her eggs to the male's brood pouch, where they are fertilized and incubated. The male seahorse eventually "gives birth" to the young, releasing them into the marine environment, where they swim and feed as biologically independent organisms.

Scientists from New Zealand's National Institute of Water and Atmospheric Research have already been successful in culturing multiple broods from wild livestock parents, which range in number from 83 to 721. The brood of 721 astonished the researchers, since it was twice the size of any previously reported brood to be released from a male seahorse of the species under study. Considering the high prices for live aquarium specimens and powdered seahorses used in Asian medicine, seahorse culture could prove to be a lucrative business, in addition to being a way to restock the natural environment. Since adults will mate readily in captivity, this eliminates the necessity for collecting brood-stock from the wild. The fact that males readily become "pregnant" again, after they release a brood of young seahorses is an added bonus. Both adults and juveniles are fed a diet of live brine shrimp and amphipods, which encourages rapid growth and optimum survival.

Increased interest in seahorse culture is demonstrated by the fact that new aquaculture operations have cropped up in Australia, Vietnam, and China. Enterprises like seahorse aquaculture have the potential to create unique local industries, which can produce seahorses which are actually superior to wild specimens—both in terms of hardiness, and in resistance to disease. This would serve to relieve fishing pressures—and ultimately, the threat of species extinction in the natural environment.[175]

"Cultural Creatives" the New Social Transformers

Since the 1960's, 50 million American adults have made profound shifts in their personal worldviews, values, and lifestyles. A core group of some 26 million represents the leading edge of this subculture; with this group differing only in its degree of intensity and commitment.

This creative, optimistic group of millions represents the leading edge of a profound cultural transformation—a change that deeply impacts their own lives, as well as society in general. Over 60 percent of this group consists of women (67% of the core group). These Cultural Creatives have an entirely new worldview and set of values. In essence, they are actively engaged in shaping a unique new culture for the 21st Century!

In the 1960's less than five percent of these Cultural Creatives were actively involved in making meaningful changes. Within a single generation, however, this percentage has grown to 26 percent! A similar poll taken by the European Union, determined that there were as many Cultural Creatives in Europe, as in the United States.

Although Cultural Creatives vary widely in age, class, and income, their common concerns are similar. These concerns include: community service, consciousness development, social justice, alternative medicine and healthcare, a "green" value system, and an enlightened vision of a sustainable future for global society.

Cultural Creatives defy characterization by personality, intelligence, religion, or ethnicity. As a group, they simply share a common worldview, a basic lifestyle philosophy, and a new set of forward-looking cultural values. Cultural Creatives regard direct personal experience as an important aspect of their lives—with some 75 percent being active in volunteer work, compared with 60 percent of the general population. Eighty-one percent of the group is vitally concerned with environmental issues, with 78 percent being in favor of finding ways to reduce our consumption of global resources. Seventy-three percent of Cultural Creatives feel that living in harmony with the Earth is important, while 68 percent believe we need to develop new, and ecologically sustainable ways of life.

The Cultural Creatives are only one of three major American subcultures. The second subculture, "The Moderns" represents about 48 percent of Americans (some 93 million). Moderns are basically typical of the entire population, with average incomes of $44,500. Moderns regard themselves as "the dominant culture." The focus of this group is on money, style, shopping, technical progress, and climbing the corporate ladder of success. Surprisingly, urban Americans have been ingrained with this "modern attitude" since before the American Revolution.

The third subculture, the "Traditionals," represents about 24 percent of American society (48 million). The group is mainly "conservative," with approximately 70 percent being religious conservatives. This group tends to favor patriarchy as dominating government, business, and family life. Their focus is on church and community, with emphasis on the customary and familiar, the regulation of sex and abortion, the freedom to bear arms, and restrictions on immigration. Many individuals of this group are pro-environmental, in the sense that they tend to be opposed to big business, and are against sweeping changes and any disruption of traditional social structure. This group is presently declining—as many young Traditionals tend to become Moderns.

Cultural Creatives can not be characterized by personality, intelligence, religion, or ethnic background. They are simply a group of individuals who share a common worldview, a set of common values and general lifestyle guidelines.[176]

**Today, something is happening
to the whole Structure of Human Consciousness.
A fresh kind of life is starting.
driven by the Forces of Love,
the fragments of the world are seeking each other,
so that the World may come into being.**

~ Pierre Teilhard de Chardin—1881-1955.

The Emerging Ecological Revolution

Most ecologists are neither liberal nor conservative. Their primary concern is to redesign virtually every aspect of our culture, and to incorporate ecologically sound principles into human lifestyles on a global scale. Essentially, they are concerned with, "Re-Imagineering" an entirely new image for society. Today's ecologists are operating at the leading edge of a truly revolutionary shift in worldview. They regard Planet Earth as a living system. Surprisingly, some 47 percent of Americans agree!

The new ecological worldview which is emerging is "Biocentric," and has a strong similarity to the worldviews of indigenous peoples everywhere. Instead of regarding our planet as a pyramid, with humans at the top, the new worldview envisions Earth as a dynamic, interactive "Living System," with humans having a major causal role in this living network.[177]

An Inexpensive, Portable Water Sterilizer

Contaminated drinking water is the world's single largest environmental killer! According to Berkeley Lawrence Laboratory Physicist Ashok Gadgil, "Four million children die annually from waterborne diseases; more than 400 deaths per hour. It's the kind of death toll that we don't jump up and down about, because it's out of sight to us."

While growing up in India, five of Gadgil's own cousins died from drinking contaminated water. Later on, he became aware that over a billion people worldwide live without the benefits of safe drinking water. Says Gadgil, "You know there's an important question that you always end up asking yourself. Where can you really make a difference? I always wanted to do something about this."

Working in his spare time, with the help of a few volunteers, Gadgil created "UV Waterworks," a portable ultraviolet water disinfection unit, that can sterilize up to four gallons of water per minute. Water disinfection with ultraviolet light is ecologically efficient, using about 20,000 times less primary energy than the common alternative of boiling water over a fire. Ultraviolet light is an effective sterilizing agent, since it deactivates the DNA in protozoa, bacteria, viruses, and other pathogens. Exposure to hi-energy ultraviolet light for about 20 seconds is sufficient to deactivate over 99 percent of the pathogens in water.

The UV Waterworks Unit, which resembles a small plastic toolbox, is equipped with a UV lamp that has a life of about 10,000 hours. The unit requires only 60 watts of electrical power, and weighs about 15 pounds. If conventional electrical power is unavailable, the unit can be powered by a 12-volt car battery, solar photovoltaic panel, or a bicycle-powered generator. The UV sterilizer, which requires an initial investment of US $300 can produce a year's supply of sterilized drinking water for a village of a thousand people, for about five cents per person. Although the impact of such a modest water purification system might at first seem insignificant, if the system were operated for 12 hours a day, approximately four million gallons of water could be disinfected over the course of a year!

The UV Waterworks Unit is currently licensed to WaterHealth International (WHI), which offers retail opportunities for local people who want to set up their own water business. The company provides the basic equipment for a small water store, for a modest set-up fee. In one village in the Philippines, UV-sterilized water sells for about half the price of name-brand bottled water. The money earned from this water store was recently used toward building a new roof for the village school.

There are over 100 Waterworks water purifiers already operating in Mexico—often with dramatic results. In the Village of Los Mogotes, for example, incidences of diarrhea dropped 93 percent after a Waterworks Unit was installed. The Philippines has nearly 100 of the devices, and there are about 100 more scattered throughout thirty different countries, including Nepal, India, and South Africa. Says inventor Gadgil, "Today 1.2 billion people don't have clean drinking water. Now, 200,000 more people are currently getting some. We ought to go from 200,000 to 200 million if we are going to make a dent. But it's a start."[178]

Since Gadgil received the original patent for his device, Water Health has experienced problems with financing and corporate structuring. Since the company emerged from bankruptcy in late 2002, its future is still uncertain. However, in addition to his teaching duties at the University of California in Berkeley, Gadgil continues to work on several other promising projects, which include a space heater for residents in the Himalayas, and a water heater for urban dwellers in India.[179]

Our Compulsion to Work

**The compulsion to work has clearly become
pathological in modern industrial societies.
Together with the obsession
of creating wealth and consuming,
it provides the impetus
to go on producing goods
at the expense of everyone's
quality of life.**

~ *Future Survey,* 24 (1): January, 2002, p. 24.

Recycling for Profit in Kenya

The Kenyan town of Korogocho is setting an example of how recycling can provide a significant source of income for the local people. In 1995 two American students, Mathew Meyer and Kenyan Benson Wikyo, established a project to recycle rubber tires from the town's dump into what they subsequently called "EcoSandals." The Project started off slowly, but continued to make steady progress. In 2001, the project put together a website, which was so successful in advertising their unique product, that $1,500 worth of EcoSandals were sold during the first three weeks!

In Kenya, a "good family wage" amounts to about one US dollar a day, so the EcoSandal enterprise has been nothing short of "miraculous" in providing a significant economic boost for this small Kenyan community.[180]

Making a Difference: The Man
Who Planted 30,000 Trees

Vishesway Dutt Saklani is a farmer from India, who lives in the foothills of the Himalayas. On January 11 in 1974, his brother was killed in a tragic accident. In his grief, Vishesway decided to create a memorial oak tree forest in his brother's memory. Each year—on the anniversary of his brother's death—Vishesway gathered his friends and relatives to plant oak tree seedlings. Since the time of his brother's death in 1974, the memorial oaks that were planted have flourished into a mature oak forest, consisting of over 30,000 trees, and covering an area equivalent to 20 football fields!

This example provides a graphic illustration of what can be accomplished by a single dedicated individual, who has the vision and determination to make a meaningful contribution to the Planetary Biosphere.[181]

Basic Strategies for Creating a Planetary Shift from Consumption-Driven to Ecologically-Appropriate Lifestyles:

1) A major shift in the consciousness of global society is needed! We must move from our overriding social focus on needless over-consumption and the accumulation of material wealth—to a new enlightened quest for spiritual wealth, happiness, and fulfillment. Once this new, consciousness has been established, consumption levels should be kept within reasonable limits of sustainability, relative to achieving a synergistic balance with the natural resources of the Global Biosphere.

2) Concurrently, global society needs to shift its main focus away from materialism and self-aggrandizement, and begin to develop the subtle-energy and spiritual aspects of the human potential. Once such a shift occurs, people everywhere will be gin to experience new levels of compassion and sensitivity for other humans and biological life forms. This new attitude will

serve to generate a renewed enthusiasm for participation in the conservation, regeneration and maintenance of Earth's natural ecosystems.

3) The biological drive for genetic reproduction should be re-directed into new channels of service, creativity, and consciousness development. Such a paradigm shift could eventually eliminate poverty and social injustice everywhere—insuring the well-being, happiness, and economic independence of present and future generations.

4) We need to abandon our wasteful, polluting activities and technologies, and make a transition to new "Green" activities and technologies. These new technologies would be designed to function within a framework of conservation, interactive energy exchange, and efficient waste-management programs which would contribute to the health and sustainability of the Global Biosphere.

5.) A major movement should be mounted, to bring creativity and art back into science, education, medicine, business, and government; moving from our egocentric positions of territorial Darwinian competition, and into new modes of cooperation and synergistic support. To be effective, this transformation needs to first occur within the collective hearts and minds of *Homo sapiens*—with the ultimate objective of eventually achieving global sustainability and social justice for everyone on Earth.

METASTRATEGY X

A WORLDWIDE SHIFT FROM A MEAT-BASED DIET, TO A PROTEIN-BASED DIET

**We have created
this overcrowded world
of overtaxed resources
by consuming ancient sunlight,
converting it into contemporary foods,
and consuming these foods to create
more human flesh.**

~ Thom Hartmann, 1998—*The Last Hours of Ancient Sunlight.*

The Environmental Cost of Eating Meat

In the United States over 30 percent of our raw materials and fossil fuels are used for raising farm animals. The energy necessary to produce a single hamburger is enough to drive a small car 20 miles! Eighty-seven percent of all the agricultural land in America is used for producing farm animals. Whereas a single quarter-acre garden can produce a year's supply of vegetarian food, the land required for a meat-based diet is approximately three acres (a 12-fold increase).

Over 50 percent of the water consumed in the U.S. is used for raising farm animals, with a meat-based diet requiring approximately 4,200 gallons of water per day! Raising food animals is also considered to be a primary cause of global deforestation, resulting in the destruction of some 125,000 square miles of forest resources every year. From an ecological perspective, every quarter-pound burger requires about 55 square feet of rainforest to produce.

The raising of farm animals is also the major cause of ground water pollution. For example, a typical egg farm with 60,000 chickens produces 165,000 pounds of excrement a week.[182] With pigs the problem becomes even more acute, since a modest pig farm of 2,000 animals produces four tons of manure and five tons of urine every day![183] When we get to cows, however, all else pales by comparison, since a single cow produces as much waste as 16 humans. On a typical feedlot with 20,000 cows, the waste produced is equivalent to a city of 320,000 people![184]

Studies at Michigan State University
have shown that it takes
2,500 gallons of water to produce
an average one-pound Steak.

~ Albert K. Bates, *Climate in Crisis,* p. 66.

The Real Cost of Shrimp

Although the dollar cost for shrimp ranges from $9 to $ 13 per pound, the real cost to the marine ecosystem is much higher. For every pound of shrimp caught, an average 5.2 pounds of by-catch is caught *and thrown back dead into the sea.* (This represents a total of 10 million tons of by-catch wastage every year!) In addition, up to 25 percent of the world's mangrove forests have already been eliminated, to make room for additional shrimp farms in coastal areas of the world.

In the Gulf of Mexico alone, *35 million red snappers are killed every year,* as by-catch in commercial shrimp fishing operations. In addition, an estimated 124,000 sea turtles are also killed—and thrown back into the sea.[185]

It's a Fish-Eat-Fish World

For each pound of farmed salmon produced, 2-5 pounds of ocean fish are required to feed them! By raising herbivorous fishes like *Tilapia,* which feed on algae, we can effectively short out the biological food chain, and thus avoid depleting our oceanic fisheries resources.

Scallops, mussels, and oysters are filter feeders. They feed on organic detritus, and photosynthetic plankton, which can harness energy directly from the sun. From an energetics standpoint, filter feeders are relatively efficient, since their position is low on the ecological food chain.

Fish and shrimp farms should not be allowed to displace natural wetlands and coastal mangrove forests, since they represent valuable ecological resources. These areas function as nursery grounds for many types of marine life. Wetlands and mangrove forests are also instrumental in protecting coastal areas from storm damage. Over the past three decades, commercial aquaculture has become an increasing and significant source of global pollution. Regulations thus need to be put in place and strictly enforced, so that waste water from aquaculture farms is properly treated before being discharged into the natural environment.[186]

Dolphin-Safe Tuna

In 1998, the National Marine Fisheries Service estimated that there were just over 2 million Dolphins in the eastern tropical Pacific Ocean. By 1999 these numbers had dropped to less than a million!

As many as seven million dolphins are estimated to have been killed before the dolphin-safe program was initiated in 1990.[187] This is just one example of how critical the timely implementation of even a single environmental program can be—relative to the impact of destructive fishing practices on our global ocean resources.

Antibiotics in U.S. Farm Animals

- **The Union of Concerned Scientists estimates that every year in the United States, nearly 25 million pounds of Antimicrobials are given to healthy pigs, chickens, and cattle for non-medical applications such as promoting growth. Over half of these (13.5 million pounds) are Antimicrobials that have been banned by the European Union for livestock production, since they are used in human medicine.**

- **By contrast, in the U. S. about 3 million pounds of Anti-microbials are used to treat diseases in humans.**

- **The term *Antimicrobial* refers to both natural and synthetic substances that target microbes. Antibiotics are a type of Antimicrobial, designed to kill bacteria.**

~ *Catalyst,* Spring, 2002 Vol. 1, No. 11, p. 19.

Green Fast-Food:
A Step in the Right Direction?

Fast Food's newest "Combo Meal" consists of an extra-lean burger, air-baked fries, and a fruit smoothie. Consumer interest in tasty fast food meals with less fat has inspired a handful of new chains from Florida to California.

Unlike the sprouts and tofu that health food restaurants served up in the 1980's, these healthier fast food meals look, smell, and taste similar to traditional fast-foods, but lack much of the junk-food baggage of their fast-food counterparts. For example, Chicago's Heartwise Express, with annual sales of $1.34 million, rivaled sales of nearby McDonalds, Wendy's, and Burger King. Their biggest sellers are beefless "Sloppy Joes," made from soy protein, and cinnamon buns made with applesauce instead of butter.

Healthy Bits Grill in Fort Lauderdale, Florida already has three restaurants, and is planning to build 29 more in the next two years. Their combo meal, which includes a 99 percent fat-free sirloin burger, baked curly fries, and lemonade, sells for less than seven dollars. "We don't sell 99 cent burgers," says the manager, "but we get kids in here who order carrot juice with their burgers."

McDonald's restaurants in Sweden have undergone an envi-ronmentally appropriate makeover. The chain now offers organic milk and beef to its customers, and about half of their restaurants now run on renewable energy.[188]

Basic Strategies for a Global Shift from Meat-Based to Protein-Based Diets:

1) Global society needs to move away from a meat-based diet—which is ecologically wasteful in terms of Earth's finite land and water resources. Instead, we should work to achieve more efficient and sustainable methods of protein production by raising free-range poultry, farming herbivorous fishes, and producing seafood using intensive aquaculture technology.

2) A concurrent transition also needs to occur to shift the focus from meat-as-the-main-course meals, to protein-as-the-main-course meals. Fish and seafood concentrates could be prepared with gourmet sauces, rice, or noodles, to create tasty, nutritionally balanced dishes, which would include delicious soups, salads, and stir-fry meals.

3) Global society needs to make a major shift away from mass-produced, nutritionally inadequate processed foods, to attractive, nutritionally superior naturally produced foods, selected for their nutritional value, and proper amounts of dietary fiber. Hormones, artificial additives, fillers, and preservatives should be eliminated.

4) A new level of social nutritional awareness needs to be established, to shift public eating habits away from fat- and sugar-intensive junk foods to tasty, nutritionally balanced meals and food items which feed the body, mind, and spirit. A coordinated global program should be organized to combat the current "Obesity Epidemic" which is sweeping across our planet. Obesity, especially in children, leads to serious health and emotional problems, and places financial burdens on the rest of society. Obesity is often a symptom of underlying nutritional and behavioral Imbalances. It also represents an embarrassing outward reflection of a person's social consciousness.

5) To adequately feed the billions of humans which presently inhabit our Planet, ecologically efficient single-cell protein should become the basic food resource for the people of Earth. A new global SuperFood program should be created to produce and

distribute highly concentrated foods derived from marine yeasts, algae, and zooplankton—which are nutritious, and low on the biological food chain. This makes them very efficient in terms of the resources required to produce them (i.e. nutrients, water, energy, growing space, and labor).

6) A shift in world consciousness needs to occur, to rekindle the public interest in the more enlightened and esoteric aspects of food, nutrition, and eating. In other words, we need to shift from a "Living-to-Eat" mindset, into a new mindset, which focuses on understanding the relationships between food and the natural environment from which it is created—and how these relationships affect our body, mind, and spirit.

METASTRATEGY XI

A GLOBAL STRATEGY FOR CREATION AND DISTRIBUTION OF SUPERFOODS

**The day that Hunger is eradicated from the earth,
there will be the greatest spiritual explosion
the world has ever known.
Humanity cannot imagine
the joy that will burst into the world
on the day of that great revolution.**

~ Frederico Garcia Lorca, 1898-1936

Food Producer-to-Table Distance Continues to Increase

A new study by the Worldwatch Institute highlights the long distances that much of the food eaten by U. S. consumers now travels. What has become apparent is that the distances between sources and destinations for American foods have increased dramatically over the past few years. According to Worldwatch scientists, America's reliance on its complex food shipping network leaves the U. S. unnecessarily vulnerable to disruptions in supply. For example, when Americans sit down to a typical dinner, many will be consuming food that has traveled between 1,500 and 2,500 miles from producer to table.

According to Worldwatch associate Brian Halweil, "The farther we ship food, the more vulnerable our food system becomes. Many major cities in the U. S. have a limited supply of food on hand. That makes these cities highly vulnerable to anything that suddenly restricts transportation, such as oil shortages, or acts of terrorism." This vulnerability is no longer limited to the United States, since food shipments have increased by 400 percent over the past four decades, during which time the global population has doubled. For example, in the UK, food now travels 50 percent farther than it did only 20 years ago.

Reliance on remote food sources tends to damage local rural economies, as farmers and local vendors struggle to compete with large supermarket chains. Long distance food shipping also opens a Pandora's Box of new possibilities for contamination, and requires the liberal use of artificial additives and preservatives to prevent spoilage. Because of the vast quantities of fuel used in food transport, the process also contributes significantly to global pollution. For example, ingredients for a typical meal purchased from a supermarket chain (including meat, grains, fruit, and vegetables), consumes from 4 to 17 times the amount of fossil fuel energy than a similar meal which uses locally produced ingredients.

It is commonly believed that the long distance food trade is efficient, since food can be purchased from the lowest-cost providers. Studies indicate, however, that regional farming communities usually gain only minimal benefits from selling their crops abroad—often suffering as a result of such international trade agreements. Says Halweil, "The benefits of food trade are a myth. The big winners are the agribusiness monopolies that ship, trade, and process food. Agricultural policies tend to favor factory farms, giant supermarkets, and long distance traders; and cheap, subsidized fossil fuels encourage long distance shipping. The big losers are the world's poor."

According to the Worldwatch study, farmers that produce food for export often go hungry themselves, while sacrificing their land and labor to feed foreign consumers. Meanwhile, poor city dwellers in developed and developing countries often find themselves in neighborhoods that lack supermarkets, local food markets, or healthy food choices. The study pointed out that, although a certain amount of foreign food trade is generally natural and beneficial, money spent on locally produced foods stays with the community longer, creates jobs, supports local farmers, and encourages local cuisines and traditional crop varieties. Halweil goes on to say, "....developing nations that emphasize greater food self-reliance can retain precious foreign exchange, and avoid the instability of international markets."

The Worldwatch Institute study questions both the wisdom and logic of long-distance food shipment; pointing out the recent massive recalls of meat products and the appearance of genetically engineered foods on the market, which have served to create a renewed interest in local foods. Consumer awareness has also generally increased, and people are beginning to understand that many supermarket fruits and vegetables are picked green and artificially ripened, in contrast to locally grown produce, which is allowed to ripen naturally on the tree or vine. For example, in the United States over half of all tomatoes are picked and shipped green—then artificially ripened at their destination.

In response to this increase in consumer awareness, restaurants, cafeterias, and large supermarkets now often offer fresh products from local farmers in their produce sections. In North America, over a dozen local food policy organizations have sprung up. These organizations help educate the public about local food options. They also lobby for local farmland protection, and help create incentives for local food-related businesses.

The Worldwatch report points out that the most powerful force behind preserving and stimulating local food markets is the consumer himself. In this regard Worldwatch offers the following basic suggestions for individual consumers: 1) Educate yourself as to which foods are in season in your geographical area, and create a diet based around them. 2) Shop at local farmer's markets, or arrange to buy from friends who produce extra food in their gardens. 3) Encourage the owners and chiefs of your favorite restaurants to purchase locally grown foods. 4) Work with community organizations to create a list of local food producers and related businesses. 5) Buy extra amounts of your favorite foods when they are in season, and freeze or can produce for year-round usage. 6) Plant your own garden and grow as much of your own food as possible.[189]

A Different Perspective of Global Food Resources

**More than enough Food is already being produced
to provide everyone in the world
with a nutritious and adequate diet—
according to the United Nations World Food Programme,
one and a half times the amount required.
Yet at least one-seventh of the world's people
(800 million people) go hungry.
About one quarter of these are children.
They starve because they do not have access
to land on which to grow food,
or do not have the money to buy food,
or do not live in a country with a state welfare system.
Genetic Engineering and Agriculture will do nothing
to address these underlying structural causes
of Hunger.
On the contrary,
they are likely to do much to exacerbate them.**

~ Jonathan Porritt, 2000, *Playing Safe:
Science and the Environment,* p. 87.

Naturally Fast-Food

Tucked away in the corner of a busy shopping mall in Bangkok, Thailand is a unique new eatery, which brings an entirely new meaning to the term "Fast Food!" The "Insects Inter Stand" is Thailand's first insect fast food restaurant, and part of a 31-kiosk network that opened in March, 2002.

Insects have always been a common traditional food item in the Thai diet—especially in the northern regions of the country. Satapol Polprapas, founder of Insect Inter, has taken the bold step of elevating the fried insect business from curbside street carts, to upscale booths in shopping malls and supermarkets.

Most of Insect Inter's patrons are young middle-class Thais, who like exotic fare, and are prepared to pay upscale prices for hygienically prepared food. Manager Polprapas states, "We buy only live insects, which are free from insecticides from our 5,000 contracted farmers. Our insects are fed with vegetables, fruits and rice."

Insect Inter's owners expect to expand their enterprise to between 100 and 150 Franchises. The owners impose strict quality control guidelines. They claim their "formula for success" is based on their own special recipe, and the unique sesame oil in which they fry their foods.[190]

In February of 2008, at a meeting organized by the United Nations Food and Agriculture Organization (FAO), experts from around the world gathered in Thailand to discuss the human consumption of insects as a food item. According to these experts, some 1400 species of insects are consumed regularly on a worldwide basis. According to the FAO, some insects in dried form have two times the protein value of raw meat and fish, while others, especially in the larval stages, are rich in fat, essential vitamins and minerals (www.freerepublic. com/focus/f-news/1973428/posts).

Spirulina: Nature's SuperFood

The microscopic blue-green algae, *Spirulina,* is one of the most concentrated forms of organic food on the Planet! It is composed of 62 to 71 percent protein, 10 percent carbohydrates, 7 percent fiber, and 9 percent minerals. It is also the highest whole-food source of vitamin B-12 available. *Spirulina* is 95 percent digestible, and contains all nine essential amino acids—plus nine non-essential ones. It is 25 times richer in beta carotene than carrots, has 28 times more iron than beef liver, and 56 times more iron than spinach or steak. *Spirulina* is rich in glycogen, and higher in chlorophyll than either alfalfa or wheat grass. For these reasons NASA has selected it as a primary high-energy food, which is ideal for space travel.

Spirulina grows at relatively high temperatures, which range from 89° to 107°F (32° to 42°C). It is very efficient in using the sun's energy via photosynthesis to convert carbon dioxide and water into a digestible, high-energy natural food. Because the cell body of *Spirulina* is not encased in a mucus membrane (as with algal forms such as *Chlorella*) it is more digestible than most other forms of micro-algae, with a digestibility rate of about 95 percent. After drying, *Spirulina* contains 62-71 percent protein, compared with 40-50 percent for *Chlorella*, 39 percent for soybeans, 18-20 percent for eggs and beef, and 7 percent for rice. It is also the world's highest source of vitamin B12—containing some 255mg per 100g (By comparison, beef liver has 80mg of B12 per 100mg).

The individual cells of *Spirulina* have a natural tendency to clump together at the surface of the water into filaments, which are about 100 times larger than *Chlorella*. This makes it especially suitable for mechanical harvesting. The taste of *Spirulina* is also superior to *Chlorella*. Although most of the *Spirulina* grown in commercial aquaculture operations throughout the world is spray dried, it can also be drum-dried, which causes it to turns brown, and taste similar to Japanese nori.

When *Spirulina* is fed to certain species of fishes, it can accelerate their sexual maturity by as much as 60 percent. It also enhances their color, and for this reason, is prized by carp breeders. It is also known to be good for dogs and cats, and it increases the shine in their coats. It is the primary food for flamingos in the wild, being one of the major factors responsible for their bright pink plumage. *Spirulina* occurs naturally in Mexico, Ethiopia, Australia, Kenya, and in hot springs, where temperatures range from 89.6° to 107.6°F (32° to 42°C). *Spirulina* was first introduced to the general public in 1827 by German Scientist, Deurban. He discovered it growing in Lake Chad, Ethiopia, where the salinity is so high that neither fish nor shellfish are able to survive. *Spirulina* is a basic food for the natives in that region, who scoop it up in special baskets, dry it in the sun, and store it in powered form. It is then mixed with wheat, baked into bread with spices, and drunk as a soup, or made into a confection. Natural populations of this blue-green algae still exist in Lakes Elementia, Rudolph, and Nakuru in Kenya; Lake Johann in Chad; Lakes Alanguardi and Circu in Ethiopia; Lake Buccacina in Peru; and Lake Texcoco in Mexico—where it has been commercially cultured for several years.[191]

The "Living Lunchbox" Concept

The "Living Lunchbox Concept" is a unique new approach for creating ultra-high-energy SuperFoods. It involves the intensive culture of small live food organisms—providing them with superior nutritional diets, so the organisms that feed on them will have higher growth and survival rates, and nutritional qualities that surpass the same species in the natural environment. Fish and shrimp produced in this way thus essentially represent "Designer SuperFoods of the Future."

Intensive aquaculture systems are ideal environments for delivering high powered nutrients to simple biological organisms such as marine yeasts, single-cell algae, rotifers, and brine shrimp. These nutrient-packeted "Living Lunchboxes" are then introduced as live food to larval and juvenile stages of shrimps, fishes, and shellfish in larger aquaculture systems.

From a perspective of biological energetics, it is very efficient to use natural Solar-Energy-Driven Photosynthesis to grow single-cell algae such as *Spirulina,* which, even in its natural state, is a "SuperFood," containing 70 percent protein, and nine essential amino acids.

Raising herbivorous fishes like *Tilapia,* that feed on algae or phytoplankton, is much more energetically and ecologically efficient than raising carnivorous fishes like trout, salmon, or catfish, which are higher on the biological food chain, and require commercially processed pelleted foods made from ocean fish stocks, thus energetically "robbing Peter to pay Paul."

Designer Diets Produce Hi-Nutrient Fish

Feeding farm-raised fish special "Designer Diets" could soon help consumers get the nutrients they need for more healthy lifestyles. According to a Purdue University research team, aquaculture fish that are fed on high fatty acid diets could provide significant nutritional benefits for the people who eat them.

The Purdue University Scientists are currently focusing their research on a type of fatty acid called Conjugated Linoleic Acid (CLA), which medical researchers have determined can be a significant factor in preventing cancer and diabetes. The National Academy of Sciences Institute of Medicine recommends that people increase their dietary intakes of foods which contain both Alpha-Linoleic Acid (an Omega-3 Fatty Acid) and Linoleic Acid (an Omega-6 Fatty Acid). They also indicated that salmon, swordfish, tuna, and shellfish, are excellent sources of Omega-3 Acids—important for building cells, nerves and eye function, and for lowering high levels of cholesterol in the blood and tissues. Both Omega-3 and Omega-6 are classified as essential fatty acids, meaning that, although they are important for human health, the body is unable to manufacture them.

In the words of researcher, Paul Brown, "We found that by adding CLA to fishes' diets, we can get more of these fatty acids into the fishes' tissues than is found in any other animal. Fish have always been the original and standard measure for good sources of Omega-3, but now we find that we can introduce other fatty acids into the fish. Next we must determine if there is an optimum ratio of Omega-3 to Omega-6 fatty acids that is healthy." Special diets for farm-raised fish can thus be formulated with nutritional additives like CLA, and grains like soybeans, to produce "Designer Fish" that contain optimum concentrations of beneficial fatty acids. To this end, the Purdue researchers are now studying different fish species, to determine just how much of these nutrients they retain when they are provided with special high-nutrient diets.

The ability to create highly nutritional "Designer Fish," (superior both in taste and nutritional quality) should provide a much-needed boost for the aquaculture industry, since maximum sustainable yields for global fishery stocks were reached at the end of the 1980's. In view of the present reduction in world fisheries resources, plus the increases in global population. Brown comments, "We have to develop new aquaculture production that rivals global production of soybeans, pigs, and chickens if we want to keep eating fish and shellfish."[192]

A Big Shrimp Story

In 1977 the author and his wife, Sharon, were invited to dinner in Miami, Florida at the home of Dr. Shao Wen Ling—an individual considered by aquaculture scientists to be "The Father of Freshwater Shrimp Culture." That evening, after enjoying a pleasant dinner served by Mrs. Ling, we adjourned to the living room for quiet after-dinner conversation. At that time Dr. Ling presented us with a piece of his traditional Chinese artwork, which depicted several giant freshwater shrimp of the commonly farmed aquaculture species, *Macrobrachium rosenbergii.* Dr. Ling also presented us with two large black-and-white photos of a new species of Super-Giant Freshwater Shrimp—a species completely unknown to the scientific world!

Dr. Ling had taken the photos two years earlier at a museum in Taiwan, where he had discovered the giant specimen preserved in a tall glass jar, tucked away in the shelves of the museum's storage area. The first of the two photos depicted the new species of shrimp by itself. The second photo showed the giant specimen, laid out on a laboratory table next to a full-grown adult specimen of *Macrobrachium rosenbergii,* the species of freshwater shrimp which is commonly farmed nearly everywhere in the tropical areas of the world. Both photos included a standard metric ruler, positioned next to the specimens, for purposes of size comparison.

The giant specimen (which we subsequently christened *Macrobrachium colossus)* had been captured in 1974 on the Island of Hainan in the South China Sea. It weighed an astonishing 623 grams (1.39 pounds)! After talking to the local fishermen in the

area, Dr. Ling discovered that these colossal shrimp were considered to be quite rare. The traditional wicker basket traps used
in the rivers on the island were designed to catch the much more
common (and smaller) *Macrobrachium rosenbergii.* Consequently,
the giant specimens were rarely captured in the standard traps. If so,
they were usually taken home by the fishermen as a prized catch
to be shared with the family. According to what the fishermen told
Dr. Ling, the preserved giant shrimp which he had discovered in
the Taiwan museum, had apparently only reached 75 percent of
its full growth-potential, as they claimed to have caught specimens
which exceeded 1,000 grams (2.2 pounds).

The implications of this important scientific discovery are that
this new undescribed species of Super-Giant Freshwater Shrimp
could theoretically have the genetic potential to grow to the size
of a 1-pount Lobster! The most practical aspect of this spectacular
"Colossal Shrimp," is that it offers possibilities for achieving up to
25 percent greater size *within the same growth period* as the commonly farmed species, *Macrobrachium rosenbergii*—in addition
to the increased marketing appeal and economic implications for
this unique scientific discovery as a new global SuperFood.

In-Vitro Cell Culture and BioReactors:
New FutureFood Technologies

A team of research scientists from Washington State University have set up arrays of experimental fermentation tanks called
"Bioreactors," containing clusters of plant and animal cells which
are suspended in nutrient solutions of salts and carbohydrates.
These cell clusters represent an innovation in biotechnology called
"Cell Factories." When these cell factories are eventually separated
from their nutrient solutions, a variety of different agricultural,
food, and medical products can be produced. The Washington
State University research team eventually hopes to upgrade their
experimental systems to a commercial-scale.

The Plant Biotechnology Laboratory in Disney's Orlando,
Florida Epcot Center is one of the major centers for Future-Food

Technologies. To date they have made a number of significant advances in plant tissue culture technology—to the point where entire plants have already been regenerated from cell clusters, or leaf-tissue sections taken from living plants. The Epcot Lab maintains a "Living Library" of about half a million tissue cultures—all growing in sterile flasks and culture vessels. This type of tissue culture technology has already been applied to strawberries, pineapples, carrots, potatoes, and peanuts.[193] Such breakthroughs in tissue culture technology could well lead to major transformations in the science of agriculture, as well as in the production of animal protein. This opens new and fascinating possibilities for what has been referred to as "Victimless Meat Production."[194]

Food technology scientist Brent Tisserat has already experimentally demonstrated the feasibility of in-vitro fruit juice production. By placing half a lemon in a laboratory vessel filled with nutrient solution, he discovered that the lemon section would produce new juice vesicles. In subsequent experiments he also found it was possible to produce juice sacs from other types of citrus fruits. These experiments open the way for learning to grow fruits (or the fruit cells themselves) without the need for either trees or orchards. In contrast, traditional agriculture requires significant amounts of time, labor, and agricultural chemicals, and displaces large areas of our natural environment with unnatural monocultures.

Microbiologist Martin Rogoff and Soil Scientist Steven Rawlins of the U.S. Department of Agriculture, envision future possibilities for creating Commercial Food Factory Systems, which would be capable of producing a wide variety of foods—including fruits, nuts, and vegetables. A visitor to such a future food factory would find no farms, no orchards, no plants, or trees. Instead, there would be vast expanses of computer-controlled laboratory culture systems, designed to produce the most desirable edible tissues of premium good crops. In order to bring laboratory scale production up to commercial scale, large amounts of "Feedstocks," in the form of essential nutrients and enzymes, would need to become available.

To this end, Biotechnologists are now working on new technologies to convert renewable materials such as wood and straw into simple sugar syrup—from which the basic elements for these feedstocks could be created.

The main advantages of food factory systems, is that they could produce food in enclosed, environmentally controlled buildings, which would be independent of local weather conditions, and would eliminate long-distance shipping costs. These advanced systems could produce optimally fresh, ripe, and nutritious food products right in major consumer centers. Other spin-off applications for this futuristic technology could produce fast-growing, disease-resistant, and heat-tolerant genetically engineered forests. We are presently on the verge of being able to genetically alter trees to speed up their growth, (as has already been demonstrated with several species of pine trees). Horticulturists at the University of Wisconsin, for example, have already successfully inserted foreign genes into embryonic white spruce trees. Many of the resulting seedlings have produced saplings that show characteristics of accelerated growth![195]

In the spring of 2001, a NASA-funded research team, led by Professor Morris Benjaminson of Touro College in Long Island, New York, embarked on a pioneering research program for growing meat (i.e. muscle tissue) in the laboratory. Ten-centimeter chunks of muscle tissue from freshly killed goldfish were placed in glass containers with a serum solution derived from cow fetuses. After a week, the researchers observed that the tissues had increased their mass by as much as 14 percent. Apparently, the partially differentiated myoblast (pre-muscle) Cells had divided to create new muscle tissue. This series of experiments clearly demonstrates the basic feasibility for growing "In Vitro" fish fillets under controlled laboratory conditions. According to Benjaminson, "The fish smelled and looked like something straight out of the supermarket. The muscle we grew in vitro looked fresh. It looked pretty darn good."

Growing larger chunks of muscle tissue under controlled conditions, presents several other problems, which have yet to be resolved. The main problem is insuring that a constant supply of nutrients reach each cell in the tissue mass. In normal tissue this occurs via a network of blood vessels and capillaries, where the capillaries are typically spaced about 200 micrometers apart. Otherwise, the tissue tends to become necrotic and dies. Benjaminson's research team has been successful in growing both white and dark muscle tissue from chickens under laboratory conditions, however, without a proper blood supply (as in living organisms) the in vitro tissue survived for only two months.

Another leading researcher in bioreactor technology is Dr. Vladimir Miranov, CEO of Cardiovascular Technologies, Inc. of Mt. Pleasant, South Carolina. Dr. Miranov feels that "Meat of the Future" will take the form of processed foods like sausage or hamburger. Instead of culturing samples of muscle tissue, Dr. Miranov is working to grow cells on protein spheres, which are suspended in a nutrient medium. These protein-tissue aggregates could then be harvested, and formed into portions to make "Patties," or "Nuggets." For the initial phase of his research Miranov will use myoblasts (cells formed at the edges of muscle fibers, which repair damaged muscles by producing new muscle cells). Since myoblasts can only survive when attached to something else, this tends to make them difficult to culture. To get around this problem, Miranov intends to mix the myoblast cells with tiny collagen protein spheres. A newly developed "Microgravity BioReactor" will create a centrifugal force that keeps the culture materials in a permanent state of free-fall. This artificial environment makes it easier for the cultured myoblast cells to cling to the collagen matrix, or "scaffold," since the system simulates the weightless conditions of space flight. Once the myoblasts have differentiated into muscle cells, they can then be harvested and shaped into serving portions.

Dr. Miranov admits that, at this time, bioreactor-produced foods are astronomically expensive. It is important to remember, however, that his prototypical synthetic hamburgers are intended to feed the space travelers of the future, so the current $10,000.-per-burger price tag fails to intimidate him. He simply points out that the first cars and TV's were also very expensive, but the advent of commercial mass-production eventually brought prices down to the point where these items were available to nearly everyone. Miranov's vision for the future is that the technology will eventually evolve to the point where bioreactors will become household appliances, so we could grow fresh sausages or hamburgers overnight, much like making bread in a home bread maker.[196]

Perspectives and Conclusions

Based on the author's 20 years of research and development experience in the fields of aquaculture and strategic food resources, it seems probable that the intensive production of FutureFoods might evolve along three major pathways:

The first futuristic food-production scenario would take the form of massive oceanic farming and processing complexes, which would combine solar energy, wave-power, and wind power to pump cold, nutrient-rich water from the ocean depths, thus creating "Artificial Upwellings." The Sun's energy would then stimulate the photosynthetic production of single-cell algae—resulting in massive blooms of phytoplankton, and the zooplankton which feed on them. Both phytoplankton and zooplankton could be selectively harvested during evening hours, using underwater lights to attract and net dense swarms of these organisms. Fertilizers, minerals and trace elements could be added to enhance the growth and nutritional value of these oceanic crops. The crop would be processed immediately, being either flash-frozen or freeze-dried, into formats which could be converted into tasty, easily prepared meal packets. These concentrated food packets could be combined with vegetables, rice, soups, or pasta, creating healthy and highly nutritious meals that could be integrated into virtually any ethnic cuisine.

A second possible scenario for FutureFood production would be to simply refine and expand existing intensive aquaculture technologies, for the commercial production of a variety of single-celled organisms such as marine yeasts *(Candida sp),* green algae like *Chlorella,* blue-green algae such as *Spirulina,* and small crustaceans such as krill (the main food source for baleen whales, which abound in the North and South Polar regions). An intensive-culture approach to food production is very efficient in terms of labor, water, space, and energy requirements—incredibly efficient, when compared to the conventional farming of chickens, pigs, and cattle. From an ecological perspective, planktonic organisms are low on the biological food chain. Thus, they are extremely "Energy-Efficient" to produce. These simple organisms offer many other advantages, in that they are small enough to be eaten whole; so there are no bones, feathers, urine, feces, or other biological waste products to pollute the environment. Finally, conventional "Livestock" requires large amounts of land, water, chemicals, fossil fuels and supplemental feeds. By contrast, vast quantities of planktonic organisms can be produced efficiently in intensive aquaculture systems, growing crops to maturity in a matter of days or weeks (rather than months, or years, as is the case with farmed animals).

A third possible FutureFood scenario would be to develop culture methods for growing entirely new, presently unknown aquaculture species, such as the super-giant freshwater shrimp, *Macrobrachium colossus,* or the giant Tasmanian crayfish, *Astacopsis gouldi,* which has been reported to reach a maximum weight of 10 pounds—and, for the first time in history, was successfully bred in captivity in 1999.[197]

Basic Strategies For the Creation and Distribution of SuperFoods:

1) A global program should be implemented to grow high-energy SuperFoods, using intensive aquaculture technology to produce high yields of concentrated nutritional food products. Futuristic food production systems could effectively reduce food-growing space by factors of a thousand-fold or more. Selected food species should include marine yeasts, phytoplankton, algae, zooplankton, rotifers, and small crustaceans such as krill.

2) Advanced hydroponics, out-of-soil farming, and intensive aquaculture technologies could be integrated into large-scale

food production systems, which could cheaply and efficiently produce hormone- and pesticide-free SuperFoods—using fertilizers which have been "Bio-Processed" from human, animal, and agricultural wastes.

3) Selective breeding and ethically responsible genetic engineering could be combined to produce SuperFoods, which would be inexpensive, fast-growing, flavorful, nutritionally superior, and suitable for any cultural diet.

4) Tissue engineering production facilities could be located near market outlets. This arrangement would create local jobs, eliminate shipping costs, and produce fresh, hormone-free, tasty, "Portion Servings," that could be fresh-frozen, or freeze-dried "on the spot," so as to eliminate contamination or spoilage, and retain maximum nutritional value.

5) International ocean sea-farming operations should be created, which pump cold, nutrient-rich water from the ocean depths to generate dense plankton blooms, which could either be harvested directly, or used to feed zooplankton, crustaceans, shellfish, or finfish. Worldwide implementation of intensive food-production systems would reduce the mounting pressures on ocean fisheries, and allow these critical natural resources the much-needed opportunity to recover to sustainable levels.

METASTRATEGY XII

NEW SCIENTIFIC
APPLICATIONS OF CONSCIOUSNESS
TECHNOLOGY

**It is remarkable that Science itself,
using its own methods,
has validated the reality
of the Four Major Psychic Phenomena.
Instances of *Telepathy, Clairvoyance,
Precognition* and *Psychokinesis*
have been verified repeatedly
in thoughtful laboratory research projects.
Further, such abilities are not rare,
but seem to be widely distributed
among the population.**

~ CharlesT. Tart, 1997, *Body, Mind, Spirit,* p. 138.

Despite the bad news regarding the negative impacts of human activities on the Global Biosphere, there are many existing and lesser-known technologies, which can provide transformative future scenarios for clean, renewable energy, and ecologically sustainable lifestyles for everyone on Earth. To achieve the practical realization of these new transformative technologies, we must first move beyond the entanglements and restrictions of 20th century thinking, and blaze adventurous new pathways into the future. Together, we must find the intellectual courage to explore the possible applications of alternative leading-edge technologies that lie outside the scientific and social paradigms of the past.

> **The main task of Psychotronics, now,**
> **is to coordinate the laws governing**
> **the Living and Inanimate Worlds**
> **and complement them**
> **with new increments of knowledge**
> **from Physics, Biology, and Psychology;**
> **this Knowledge will derive**
> **from specific manifestations**
> **of the human psyche,**
> **with Man or an appropriate model**
> **serving as the Interlink.**
>
> ~ John White & S. Kripper, Eds., 1977- *Future Science,* p. 48.

Radionics: An Alternative Subtle-Energy Technology For the Third Millennium

The Science of Radionics (i.e. Psychotronics) deals with interactions between matter, energy, and consciousness. It is based on the research of Dr. Albert Abrams. In the 1940's Dr. Abrams and several subsequent researchers developed a series of unique electronic instruments, which apparently created a subtle-energy linkage between doctor and patient. These devices contained the following basic components: 1) a radio-frequency amplifier, 2) a variable condenser tuning system, and 3) non-inductive resistance, which occurred on a bakelite "Rubbing Plate."

A sample of the patient's blood, saliva, hair, or urine (and more recently, a Polaroid photograph of the patient) was placed in the "Tuning Well" of the instrument (a glass beaker, with a coil of copper wire wrapped around the outside). When the operator's fingers encountered resistance on the bakelite rubbing plate, specific numerical values could be established, using sets of calibrated dials. Once the numerical value set for a specific illness had been established, the polarity of the instrument was reversed, and it began to "Broadcast" a particular set of healing frequencies, which apparently functioned to cancel out the patient's illness.

Radionics (much like Tom Bearden's MEG device) apparently operates outside the normal space-time coordinates, since healing has been routinely achieved even when the doctor and patient are physically separated by thousands of miles. Although many patients considered "Terminal" by conventional medical doctors were cured by this technology (with hundreds of such cases being well-documented in the medical literature), the American Medical Association eventually banned the use of medical radionics in the United States.[198, 199]

In Great Britain, and some other countries outside the United States, radionics is professionally certified as a legitimate medical practice. Successes with radionics are also well documented in U.S. Scientific Literature in Agriculture, and Veterinary Medicine (which were *not* banned). Since the invention of the original radionics devices in the 1940's, the technology has quietly been upgraded from its original vacuum-tube design, into a solid-state electronic format. Within the last two decades, radionics devices have been combined with computer technology, and now often come with sophisticated software for medical diagnosis and treatment.

It is the author's hypothesis that entirely new fields of 21st century science could be created by combining radionics with remote satellite sensing technologies. There is good reason to believe that an entirely new science could be created for the purpose of monitoring and repairing Earth's major environmental problems—on a planetary scale. This basic concept could also be extended to introduce new and positive paradigms into the global social consciousness; thus opening exciting possibilities for creating positive future scenarios for peace, social stability, and environmental sustainability for the present and future generations of human society.

The "Eyes" of Gaia

The space around Earth is filled with large numbers of satellites that record and transmit data back to groundside computers. It is estimated that there are well over 2,000 satellites currently in space.

Paris-based Earth Observation Technology, which began over 26 years ago, has nearly 300,000 employees from its 15 member nations. This agency continues actively developing satellite launch vehicles and scientific spacecraft, as well as communications satellites and vehicles for manned space flight. One of this agency's most important functions is climate monitoring. For example, Earth Watch Satellites, using synthetic aperture radar (SAR), can penetrate cloud cover to detect changes in vegetation, land use, and Earth surface temperatures. Other remote-sensing satellites such as "ERS" and the eight-ton "Envirosat" can measure changes in sea level, sea surface temperatures, plankton blooms, ocean winds, and thickness of the Greenland and Antarctic Ice Sheets. Longer range objectives for these Earth-sensing satellites include the monitoring of environmental destruction, weather modification, and implementation of international environmental treaties. As new sensing technologies are developed, they will be incorporated into succeeding generations of these Earth-sensing satellites.

Other types of "Eyes" include: weather satellites, The International Space Station, upgrades for the Hubble Space Telescope, an Earth-orbital ultraviolet telescope called GALEA, and satellites launched into outer space to study various aspects of the Sun, Moon, Mars, Mercury, Saturn, asteroids, and comets.[200]

From a consciousness technology perspective, as sensory data continues to be downloaded from the satellites and fed into the computer nodes of the global communications network; the planetary awareness, (i.e. global brain) should evolve correspondingly. In addition to the expansion of the planetary consciousness, a growing awareness of our solar system (solar consciousness), and eventually the galaxy (galactic consciousness) will occur, and within an increasingly accelerated time-frame. Continuing along this line of thinking, it becomes apparent that the recent extraordinary expansion of our planetary self-awareness over the past few decades should also be reflected in correspondingly profound shifts in human awareness, as well as in the collective consciousness of the entire human race.

Psi-Microvision: A Psi-Tech Tool for Re-Inventing Traditional Science and Technology

To date, physicists have defined some 200 subatomic particles, using the sophisticated technologies of high-energy physics. Unknown to most of conventional scientists, however, is the fact that detailed observations of molecular, atomic, and subatomic structures were made nearly 80 years ago!

A form of psychic perception, known as "Magnifying Clairvoyance," or "Micro-Psi," has existed for thousands of years as hidden knowledge, known only to Eastern adepts. In its classical format, the technique does not involve magnifying the object under study, but in "making oneself infinitesimally small," thus enabling microscopic particles to be studies in minute detail.

In 1895, Theosophical Society members, C. W. Leadbeater and Annie Besant, began an intensive research program, which continued for 38 years. Beginning with simple elements like hydrogen, nitrogen, and oxygen, the investigators eventually included all the known elements of the times, plus selected organic and inorganic compounds.

Using micro-psi technology, the object under study was described to an assistant, who then created detailed sketches. Atoms were depicted as highly structured bodies with definitive external structures, and interiors which were subdivided into smaller compartments of spherical, ovoid, conical, and other shapes. These bodies were divided into even smaller compartments, which were, in turn, observed to contain even smaller components. Since the atoms under observation were vibrating at extremely high frequencies, the researchers found it necessary to develop psychokinetic Techniques for slowing this motion down—thus enabling them to make accurate observations and counts.

In 1979, Cambridge University physicist, Dr. Stephen Phillips, developed an improved version of Classical Quark Theory, which subdivided each quark into sub quarks, or omegons. The discrepancies between Current Physics Theory and Dr. Phillip's New Theory were bridged by reinterpretation the Micro-Psi Data of Leadbeater and Besant; who 85 years earlier, had claimed to actually "see" the atoms exactly as they were described in Dr. Phillip's 1979 Scientific Paper, which appeared in the Journal, *Physics Letters*.[201]

The "Force-Multiplier Effect" An Effective Tool For Cultural Transformation

A transformative idea or concept is first expressed in written, audio, visual, or digital format. This "Information Packet" is then distributed over the global Internet via E-mail. Each recipient of this message is then asked to pass this "Information Meme" along to his or her acquaintances and professional associates, asking them in turn to send the information on to their own friends and professional associates.

Through the various electronic, digital, and organic transmission modalities, the force of the original message is multiplied exponentially—as it spreads fractally throughout the global data-sphere. The Internet might thus be said to possess many of the characteristics of a living entity, since it has the innate capability to grow and evolve, according to its associated input and technologies. The Internet thus represents an extremely powerful "Force Magnifier," which can fractally amplify the power and influence of individuals and organizations and transformative new concepts throughout the electronic global "MindField."

Collective Consciousness: A Powerful Transformative Force

The power of collective consciousness has tremendous potential for bringing meaningful and lasting change to world society and the Global Biosphere. In a study recently conducted by Dr. Mitchell Krucoff, head of the cardiovascular program at the

Duke University Medical Center, a number of patients complaining of cardiac chest pain was divided into two groups of equal size. One group was "prayed for," and the other group was "not prayed for." The request for prayer was accomplished through the aid of a global prayer group, which was already organized on the Internet. The interdenominational group included Buddhists from Nepal and Tibet, Hindus in India, some Carmelite Nuns near Baltimore, a prayer group from Jerusalem, some Unity Church members, and a group of fundamentalist Protestants from North Carolina.

Results from the initial study were stunning! Patients in the designated "prayed for" group were observed to have between 50 and 100 percent fewer side effects than the control group that was not prayed for. The study was subsequently expanded to include twelve other major U.S. hospitals. Preliminary data from this study is slated to be published in the *American Heart Journal.*

An interesting sidelight of this study is that the positive results with human subjects are supported by dozens of other studies which used non-human subjects! This essentially counteracts a major argument of traditional scientists, who often cite the possibility of "hidden biases" surfacing in research results, due to "conscious volition" in human patients (i.e. "the Power of Positive Thinking").

With animals, plants, bacteria, and germinating seeds, this "conscious bias factor" essentially ceases to exist. Such studies included healing rates of surgical wounds in animals, bacterial replication rates, growth of fungal colonies in laboratory petri dishes, growth rates of seedlings under controlled laboratory conditions, and the rates of specific biochemical reactions in test tubes. Such experiments on non-human subjects can be conducted under precise laboratory conditions—thus effectively eliminating most of the major objections cited by skeptics of human studies in areas of conscious research.

Another fascinating aspect of this type of research is that the non-human studies highlight the fact that "Nonlocal Consciousness Effects" can operate across a broad spectrum of natural phenomena to yield measurable effects ranging from the "Micro

Realm" (i.e. atomic and molecular levels), to the levels of bacteria and larger single-celled organisms, and extending up through the levels of multicellular organisms, which include invertebrates and vertebrates. In the words of Larry Dossey, M.D., "This lineage, this so-called concatenation or coming-together of effects unifying these vastly different domains of nature, is one of the most compelling aspects in the field." The significance of this "Unifying Concatenation" represents a highly valued aspect of valid scientific theory and practice, and suggests that we are dealing with a general principle that is embedded everywhere throughout the patterns of nature.

The studies have focused on two main biological processes: healing and fertility. The healing research involves wound repair, as well as the healing of specific diseases like AIDS and coronary artery disease. Fertility research has involved humans, as well as nonhumans—mainly the growth rates of plants, and germination rates of seeds. In the words of Dr. Dossey, "The secular scientific and the religious sides of life cannot be kept in separate boxes." Notable scholars like Alfred North Whitehead have expressed similar sentiments with regard to the inseparability of science and religion as follows: "It is no exaggeration to say that the future course of history depends on the decisions made by this generation as to the relationship between science and religion." Ralph Waldo Emerson said it even more dramatically: "The religion that is afraid of science dishonors God, and commits suicide." In Albert Einstein's words, "Science without religion is lame. Religion without science is blind."

Research of this type would thus seem to emphasize the importance of both religious tolerance and interfaith cooperation, since regardless of the religion, the end results appear to be truly universal. Studies such as these would thus seem to provide a great measure of hope for our presently troubled society and our environmentally stressed biosphere, since they show that the old-fashioned concepts of love, empathy, and compassion have the power to change the state of living organisms, and their natural environment.[202]

Internet-Based Torsionic Meditation:
A Mechanism for Shifting Global Consciousness

In the 1970's a U.S. Air Force Group in Thailand donated some obsolete computer equipment to a local community of Buddhist monks, who began experimenting with the computer hard drives—attempting to use them as modern techno-analogs of traditional Buddhist prayer wheels.

According to leading-edge consciousness researchers in Russia, the groundwork for Torsionic Meditation was established by spiritual teachers Nicholas and Helen Roerich. From a consciousness research perspective, Buddhist prayer wheels and computer hard drives can both be regarded as generators that can propagate faster-than-light torsionic emissions—said to be capable of carrying messages over interstellar distances.

Computer hard drives, are essentially finely-tuned, spinning fractal multi-magnets. Since they are mass produced, they offer the advantages of being relatively inexpensive, yet are created within precise manufacturing specifications. With nearly half a million PC's in the world today (over 100 million of them connected to the global Internet), the potential exists for a "Global-Scale Phased Torsionic Antenna," that could be effective for purposes of collective meditation, or global consciousness paradigm shifting.

Unique strategies for consciousness upliftment and planetary defense (SDI—the "Spiritual Defense Initiative") were originally suggested by such spiritual teachers as Swami Satchidananda, who was influential in the formation of the still active Pentagon Meditation Club in Washington, DC; and the Russian Initiative Group for Defense of Earth, whose members are skilled in the scientific exploration and research of "Inner Space."[203]

Basic Strategies for New Scientific Applications of Consciousness Technology:

1) A new science of Earth-energy technology should be developed, to integrate the principles which relate to the global

energy grid, ley lines, and Telluric power points with techno-logically updated applications of "Energy Architecture" and "Geo-puncture." This new science could be used to create subtle-energy fields which would enhance human health and intelligence, improve the health and nutritional value of food crops, and help restore and revitalize the global ecosystems.

2) An International Research Program should be established to explore the scientific applications of group-consciousness technology, possible applications of field-consciousness effects, and creation of a technologically amplified "Global Consciousness Internet," for enhancing human consciousness and repairing ecological damage to the Global Biosphere.

3) A New International Scientific Program should be initiated to explore subtle-energy technologies, with the objective of de-veloping new scientific applications for the science of radionics, relative to repairing ecological damage from human activities, and boosting human consciousness to new levels of functional efficiency.

4) Responsible applications of genetic engineering and re-lated technologies should be used to create superior varieties of organisms, which would be fast-growing and more resistant to disease and the environmental stress factors which result from pollution and global climate fluctuations.

5) An International Research Program should be established, to integrate consciousness technology with quantum physics. By combining psi-microvision with physics, chemistry, biology, and medicine, entirely new concepts would emerge—in virtually all areas of science and technology.

6) New applications of remote viewing technology should be developed, to shift the focus from "Military Intelligence" to "Scientific Intelligence." Such a shift would allow scientists to gain new insights into the nature of life itself, and could provide a cost-effective means for the preliminary exploration of outer space. This would save millions of dollars in research funds, and help to avoid many of the dangers of space travel. Military applications of remote viewing represent only "the tip of the

iceberg" with regard to its other exciting potential applications, such as "Consequence Analysis," (i.e. evaluating the possible success or failure probabilities for any action which is initiated in the present).

7) The infant science of "Future Memory" provides another example of a developing consciousness technology which should be brought into the scientific mainstream for further refinement, with the objective of creating another effective system for evaluating future scenarios; thus saving valuable time, and countless millions of research dollars, which could otherwise be used more effectively for creating successful outcomes and products.[204]

THE FUTURE OF THE GLOBAL BIOSPHERE: PERSPECTIVES AND POSSIBILITIES

An Alternative Worldview is taking shape,
championed by scientists, environmentalists,
relief workers and human-development experts.
We must cast off the 19th Century Mentality
that structures the current system, and develop Innovative Models
of Sustainability,
based on recognizing
the Interdependence of all Life
and committed to Conserving Resources
for Future Generations.

~ Juliet B. Shor & B. Taylor, *Sustainable Planet: Solutions for the Twenty-First Century* (From Review by *Future Survey*, 25:1, January 2003, p. 18).

Despite the many impending dangers which presently threaten the socio-ecological fabric of planet Earth, there are new discoveries, some good news, and new approaches to thinking and action, which illustrate the fact that it *is* possible to reverse the present trends of environmental degradation, and transform our Society and the Global Biosphere in ways which, until recently, seemed impossible. A sampling of these positive new discoveries and ideas is presented in the sections which follow.

A Promising Recovery for the Great Barrier Reef

According to a report released recently by the Australian Institute of Marine Science, the Great Barrier Reef has made a significant recovery from severe coral bleaching, overfishing, pollution, sedimentary runoff, and the impacts of coastal development. The report states that, during the past few decades, nearly 60 percent of the reef had been affected by coral bleaching, a condition that occurs when environmental stress causes the symbiotic algae that inhabits the living coral polyps dies off—causing the coral to lose its color and turn a ghostly white.

According to Dr. Clive Wilkinson, Chief Scientist for the International Coral Reef Monitoring Network, "Reefs if left alone and not stressed, will recover quite rapidly." The report praised the Australian government's Great Barrier Marine Park Authority for its effective role in reef restoration by setting up marine sanctuaries, protecting fish stocks, and monitoring and maintaining effective water quality standards. The 1200-mile-long Great Barrier Reef off the east coast of Australia constitutes the largest complex of islands and coral reefs in the world. It is comprised of over 2,600 individual reefs, and 300 islands. The Great Barrier Reef is also Australia's most popular tourist attraction.

Dr. Wilkinson feels that many of Earth's coral reefs are still "on the cusp" between survival and extinction, due to the effects of severe coral bleaching during recent years. He is, however, optimistic in his outlook for the future of the global coral reef ecosystems, expressing the opinion that most of these reefs will eventually recover, because of improved marine management policies imposed by world governments. "I think we're about to see major improvements in coral reefs," he said. Dr. Wilkinson also expressed the opinion that the main danger to the world's coral reefs in the future, is from the effects of global warming.[205]

Africa's Hidden Waters

UNESCO scientists recently announced the discovery of vast water resources located deep below the deserts of Africa. These "Environmental Treasure Troves" represent precious ecological assets in the impending water crisis which the inhabitants of this region face.

At a recent meeting of geohydrologists in Tripoli, The First Continental Survey of Hidden Underground Aquifers was presented. Aquifers, much like rivers, often cross international boundaries, but unlike rivers, there have as yet been no international regulations for sharing aquifers.

In addition to the vast Nubian Sandstone Aquifer, which lies below Libya, Egypt, Chad and Sudan, UNESCO Scientists also identified at least 20 other transnational aquifers on the continent of Africa. The group expressed its concern that population increases in these African nations will likely result in growing tensions over water, as these nations scramble to pump out as much water as possible for their own use.[206]

"Flashing Fish" BioMonitor

Scientists at Singapore National University have created something so bizarre that it borders on the realm of science fiction: a transgenic zebra fish that flashes red and green when exposed to toxic chemicals. Biomonitoring fishes and other species of living organisms offer intriguing possibilities for use as "Organic Instruments" to monitor the water quality in aquatic environments. The high degree of sensitivity of certain species of fishes regarding their abilities to detect minute traces of already common scientific knowledge, since fish like salmon can detect minute chemical differences in water quality, in concentrations as dilute as parts per billion!

Other scientists, however, have already expressed their concern about releasing such transgenic fish into the environment, suggesting that the ecological impact of such introductions could range from "inconsequential," to the complete extinction of nontransgenic natural populations of the same species. Since zebra fish rely on color for their mating displays, it is possible that these newly bioengineered "Flashers" might gain a reproductive advantage over their natural counterparts. Whatever the case, considering the worldwide demand for exotic and novel aquarium specimens, it should not be long before we begin to see even more bizarre bioengineered creatures swimming around in home and office aquariums.[207]

An Elephant Orchestra

At the Thailand Elephant Conservation Center, the residents earn their keep by giving rides, demonstrating traditional logging skills, and painting pictures for the tourists. Since Elephant's hearing is much more sensitive than their sight, they use a wide range of vocal signals to communicate. This thus makes them ideal candidates for music-making.

American elephant expert, Richard Lair, and his friend, neurochemist/musician, Dave Soldier, hatched out the idea for an elephant orchestra several years ago in New York City. They worked together to develop special "Jumbo-Sized" versions of traditional Thai musical instruments such as tubular bells, a xylophone-like renat, an African drum, a thunder sheet, harmonicas, and a gong created from a circular saw blade, which had been confiscated from illegal loggers.

The "Group" has already produced its first music CD, and is working on a second. Conductor Lair worked out a set of hand signals for the handlers to cue the elephants to play. It is hoped that the new CD's will raise public awareness of the tragic state of Thailand's elephant population, which has dwindled to 2,500 from 100,000 only a century ago.

Lair continues working to develop new elephant instruments, and is also actively seeking new talent. One three-year-old in the region has already proved to be a rising pachyderm prodigy, and another elephant center in Thailand is working to develop its own Orchestra.[208]

Return of the Swordfish

A fish stock assessment conducted in Spain for the International Commission for the Conservation of Atlantic Tunas (ICCAT) indicated that North Atlantic swordfish stocks have increased from a low of 65 percent, back to approximately 94 percent of their historical optimal levels. This indicates that the depleted swordfish populations have responded positively to strict catch limits, and are making a swift recovery back to previous population levels.

According to Bill Hogarth, Director of the National Marine Fisheries Service, "This is good news for those who care about the long term health of the North Atlantic swordfish." The US/ICCAT delegation was an alliance between commercial, recreational, and conservation groups, who had combined their efforts to rebuild this formerly lucrative sport and commercial fishery.

When the swordfish have recovered to 100 percent of what 1CCAT considers to be a healthy stock level, American fishing vessels will once again be able to make a healthy profit. Says Hogarth, "Commercial and Recreational Fishermen in the United States supported the strong management action ICCAT implemented four years ago, and once the stock is fully rebuilt to the ICCAT goal, we will push for a quota that rewards them for their sacrifices."

The Bronx Zoo's Wildlife Conservation Society (WCS) began working to protect the Swordfish Populations in 1999. According to Dr. Ellen Pikitch, Director of Marine Programs for WCS, "The recovery of the North Atlantic swordfish population clearly shows that good science can go hand-in-hand with good fisheries management. This is an incredible victory for conservationists, commercial fishermen, consumers, and of course, swordfish."

Besides imposing strict catch quotas, certain areas were closed entirely by the U.S. to protect the juvenile swordfish, and thus help speed the rebuilding of swordfish stocks. At the October 2002 Meetings of ICCAT in Madrid, WCS Scientist, Dr. Beth Babcock, announced that the stocks were rebuilding much more rapidly than had been expected. She stated, "If recent catch levels continue, there is more than an 80 percent chance that the population will recover by 2009, or even sooner." During the 1970's sword fishing was a popular sport for recreational fishermen in the U.S. As a result, North Atlantic swordfish stocks were over-fished. Recent focus on the sport has created a renewed interest in recreational swordfishing tournaments.

Many conservation organizations have worked together to save the North American swordfish. Additional support was provided by the "Give the Swordfish a Break" campaign, a consumer awareness project, which combined the efforts of Sea Web and the National Resources Defense Council. This campaign was designed to influence the eating habits of restaurant clientele. In January of 1998, the "Give Swordfish a Break" campaign was launched in New York, where 27 east coast chiefs announced that they were removing North American swordfish from their menus, until a recovery plan was in place. Eleven months later, over 500 more chiefs nationwide had responded in kind.

In conjunction with this type of consumer awareness campaign, the Audubon Society's Living Ocean Program plans to release a "Seafood Awareness Card," that will provide eco-conscious consumers with information on the conservation status of many different types of commercial finfish and shellfish. Meanwhile, ICCAT continues its efforts to expand fishing quotas to include other endangered species on a global scale. Proposed regulations include: recreational permits, dockside reporting for recreational catches, and recreational bag limits.

Fortunately, swordfish grow rapidly, and females begin reproducing when they are only five years old. Spawning normally occurs in warm tropical and sub-tropical waters. Commercially, swordfish are caught with deep-ocean longlines, at night. In the words of Dr. Pikitch, "While the recovery of North Atlantic swordfish populations is a great victory, the battle is far from over. We need to continue to protect juvenile swordfish, and minimize by-catch of severely depleted species such as white and blue marlin, which are still accidentally taken by swordfish boats." Dr. Pikitch also pointed out that the status of swordfish in the South Atlantic and Pacific regions is presently unknown, but is also probably overfished.[209]

Antarctic Ozone Hole Closing?

In a recent United Nations report, a leading atmospheric scientist claims that the hole in the ozone layer over Antarctica may close as early as the year 2050. According to Paul Fraser, scientific researcher for CSIRO (Australia's Commonwealth Scientific and Industrial Research Organization), "The major culprits in the production of the Ozone Hole are CFC's, and they have started to decline in the atmosphere. We think the Ozone Hole will recover by about 2050."

Chlorine from CFC's (Chlorofluorocarbons) is the main chemical culprit responsible for destroying the ozone layer over Antarctica. CFC's, used since the 1930's in refrigerators and air conditioning units, tend to remain in the atmosphere for decades. The recent United Nations report indicated that concentrations of ozone-depleting gases in the lower atmosphere have been declining steadily since the year 2000, and that slow but steady progress was being made toward recovery of the ozone layer.

Under the 1987 Montreal Protocol, developing countries committed to reduce the production and consumption of CFC's by 50 percent by the year 2005, and to achieve a cut of 85 percent by the year 2007. Scientists monitoring atmospheric CFC concentrations on Australia's southern island of Tasmania indicated that in 1950 atmospheric concentrations of CFC's began rising from zero, to a peak of 2.15 parts per million by the year 2000. Says Fraser, "We are now at a point where the atmosphere can actually remove CFC's faster than they are being re-introduced into the atmosphere." The UN Report, authored by Fraser, indicates that this reduction of atmospheric CFC's is proof that the Montreal Protocol was effective. Although the scientific data which was presented confirmed the slow decline in ozone-depleting gases in the lower atmosphere, it also warned that the ozone layer would remain vulnerable to pollution from CFC's and similar airborne chemicals.

From an environmental perspective, the ozone layer is essential for life on Earth, since it screens out harmful ultraviolet radiation from the Earth's surface. The report also warned that the Antarctic ozone hole would only close completely, if the nations of the world continued to adhere to the Montreal Protocol Guidelines, and if no additional factors negatively impacted the ozone layer (such as major increases in greenhouse gases). The report concluded stating, "These results confirm that the Montreal

Protocol is achieving its objectives. During the next few decades we should see a recovery in the ozone layer."[210]

Cloning Earth's Oldest Trees: A Priceless Legacy for Future Generations

A government agency and a non-profit organization are combining forces in an effort to clone the world's oldest tree—a bristlecone pine tree, located in the California high country. A partnership between the U.S. Forest Service and the Champion Tree Project International is attempting to clone a 4,767-year-old bristlecone pine tree, affectionately known as "Methuselah."

For centuries, bristlecone pines have flourished in the higher elevations of the Great Basin. They range from California to Colorado, and are distinguished by their ability to endure harsh environmental extremes, and to adjust to changing environments. Until 1953, their legendary age was not commonly understood, until Edmund Schulman discovered a bristlecone pine in the White Mountains of the Inyo National Forest which had been dubbed, "Patriarch" by a local forest ranger. Further study of this 36-foot-wide tree, indicated that it was 1500 years old. Subsequently, even older trees were discovered, with the oldest trees being found at elevations between 10,000 and 11,000 feet.

The most ancient bristlecones all showed extensive areas of dead wood, with only thin strips of living bark. Surprisingly, those trees found to be growing under the most extreme conditions, with little soil and moisture, turned out to be the oldest. As the study progressed, several older trees were discovered that proved to be from 3,000 to 4,000 years old. The first tree that exceeded 4,000 years old, Schulman named "Pine Alpha." In 1957 an ancient bristlecone pine was found, which was dated at 4,723 years old. This tree (since named "Methuselah") has now achieved a famous place in history, as Earth's "Oldest Living Tree." Says Schulman, "The capacity of these trees to live so incredibly long may, when we come to understand it fully, perhaps serve as a guidepost on the road to understanding longevity in general." An understanding of the genetic constitution of these ancient trees will hopefully provide scientists and foresters with the knowledge and tools to enhance the health and longevity of trees everywhere on Earth.

Jared Milarch, the Co-Founder of the Champion Tree Project International, has cloned many champion trees, which have been planted throughout North America. He explains, "What we are trying to do is leave a living legacy that future generations can enjoy, and preserve the last remaining old growth forest genetics that we've got. We can preserve that link to the past and show future generations what it used to be like."[211]

Traditional 20th Century Paradigms

Prior to the Year 2000, paradigms typically took the form of mental matrices, behavioral models, thought-forms, or operational protocols for thinking and action. For leading-edge 21st century thinkers, these old paradigms come across as restrictive thought patterns, which are essentially societal hangovers from the Industrial Age.

Although the following quotations were made by some of the most respected scientific and military authorities of their times—a generation or two later these "Truths" have become caricatures of the times in which they existed. They now seem almost stranger than fiction!

X-rays are a hoax.

Aircraft flight is impossible.

Radio has no future.

~ Lord Kelvin, Physicist and Mathematician—1824-1907.

Heavier-than-air flying machines are impossible.

~ Lord Kelvin, President, Royal Society of London, 1895.

> ## The bomb will never go off,
> ## and I speak as an expert on explosives.
>
> ~ Admiral Wm. Leahy, U.S. Atomic Energy Project.

Transitional or Gateway Paradigms

Transitional paradigms are catalytic, provocative, and revolutionary, when compared with the mainstream thinking patterns of the times. These new futuristic paradigms represent "Metaphorical Gateways" for the mind and consciousness to explore entirely new and novel possibilities for thinking and action. The following quotation by astrophysicist, Carl Sagan, is an example of a transitional paradigm. It gives an unexpected and provocative "tweak" to a traditional American saying:

> ## If you wish to make Apple Pie
> ## truly from scratch,
> ## you must first invent the Universe.
>
> ~ Carl Sagan—1934-1996

New Paradigms for the TriMillennium

New paradigm models for the Third Millennium would differ markedly from the restrictive paradigms of the past. Using sophisticated computer technology, new paradigm models could be created to function as "Self-Aware Concepts," which would have the innate capabilities for self-organization. They would also be endowed with the abilities to grow, change, and evolve in ways which would provide the most beneficial support for the social and ecological parameters of the times.

Unlike the old paradigms, which were severely rigid and constrictive, new millennium paradigms would have flexible and evolvable guidelines built-in. These guidelines would function to enrich mental and spiritual growth, encourage intellectual diversity, and help us develop new and creative solutions for the social, economic, and environmental problems which we presently face as citizens of planet Earth.

Embedded in the conceptual matrix of this new paradigm model would be a constellation of new solutions, technologies, concepts, and programs, which would include the basic idea of "Designing for Posterity." Each proposed decision and action would thus be evaluated in terms of its potential impact on the Future. TriMillennial Concepts and Programs would include such concepts such as "Giving back more than you take," and "Leaving a place a little nicer than you found it." This simplistic type of positive thinking—when multiplied geometrically by the millions of human minds on Earth—would thus function to create a wide selection of bright future scenarios for both the global environment, and for the generations of humans to come.

**When a critical number
of the world's population
reaches a state of spiritual vibrations,
our institutions will change.
Institutions are created
to facilitate, regulate, and guide
Human Behavior.**

~ Valerie V. Hunt, *Infinite Mind,* p. 295.

A New TriMillennium Center for Future-Science Technology and Planetary Enhancement

It is the author's conviction that global leaders and motivated individuals everywhere must work together to halt and reverse the de-evolutionary forces which presently threaten the fabric of human society and the Global Biosphere. Within this context, a Global Foundation for Future-Science Technology and Environmental Restoration should be established. This new planetary center (The TriMillennial Center) would be created to develop and implement alternative new technologies, and to create future scenarios which would transform society, and human consciousness itself. In practical terms, this Supernational Organization would develop enlightened programs in areas of international cooperation, conflict resolution, social justice, and ecological sustainability. The Developmental Plan for this New TriMillennial Center might take shape as follows:

The first step in creating the TriMillennial Center would be to assemble a core group of visionary leaders from industry, government, education, the creative arts, and NGO's for the purpose of producing a set of basic guidelines and objectives. This core group should be capable of generating its own basic funding, which would serve to reduce the self-serving and disproportionate influences of government, commercial, and military interests.

The focus of the TriMillennium Project would be to research, develop and implement, clean, ecologically appropriate technologies, which could be integrated into our existing social and economic infrastructures, thus upgrading the fabric of human society in areas of transportation, communications, power-generation, food production, manufacturing, marketing, packaging, waste processing, and housing. This new vision for a sustainable society would function "in compliance" with the laws of nature, and for the mutual benefit of all life on Earth.

In accordance with the basic guidelines and objectives, the founding group would elect a TriMillennial Council, which would convene regularly via virtual conferencing, to create and implement a new design for global society. Their ultimate objective would be to achieve social justice for every person on Earth, while moving human social consciousness into a framework, appropriate for survival and success in the Third Millennium.

The first "TriMillennium Center" might initially take shape as an International Conceptualization, Planning, and Conference Center, built from the ground up, to form a nurturing, synergistic, and enhanced environment, where participants could live, work, and interact for periods ranging from several days, to months at a time.

This TriMillennium Center should be strategically located for convenient access via international air transport. The Center would be equipped with state-of-the-art telecommunications, in order to encourage virtual interactions with centers and individuals in other locations. Similar centers could eventually be established in strategic locations around the world. Although start-up funding assistance could be provided, these satellite centers should employ the human resources they would attract, to rapidly become financially self-generating, applying the basic design concept of sustainable futuristic eco-resorts, created in the spirit of the sustainable cities, or integrated island communities envisioned by futuristic architects like Doug Michels and Jacque Fresco. By seamlessly integrating human space with the natural environment, the atmosphere thus created would enhance health, vitality, happiness, creativity, and intercultural social exchange. In addition to providing comfortable living and working spaces, TriMillennial Centers would offer an array of recreational activities, which would relate directly to the unique characteristics of the surrounding environment.

The TriMillennial Centers might initially take the form of a series of conferences, which could be held at resorts located in different parts of the world. Teleconferencing facilities would serve to encourage participation of individuals living in remote areas, islands and wilderness regions. As the TriMillennial Center concept grows and evolves, certain appropriate locations could be developed into high-tech futuristic eco-villages and eventually eco-cities, which would be designed for maximum human health and comfort, while protecting and enriching the surrounding environment.

For leading professionals and cultural creatives, long-term housing and retirement villages could be established at selected locations, to encourage mentorship programs for grooming the younger generations for future leadership roles. Eventually a network of compact satellite centers, established at strategic locations, would serve to represent the "Nodal Points" of a greater "Global Consciousness Net." This evolutionary information-communications resource would thus eventually be accessible to anyone, anywhere on Earth.

Final Thoughts

If these Twelve Meta-Strategies can be successfully integrated into the biological and socioeconomic fabric of planet Earth, humans can successfully complete their co-evolutionary journey into the Third Millennium and beyond—instilled with a renewed spirit of enthusiasm, enlightened accomplishment, and joyful satisfaction, for their new symbiotic relationship with planet Earth and their fellow humans. As our species continues to evolve, *Homo sapiens* might conceivably re-create itself into an entirely new format—a new species called *Homo novus*. The possibilities for such an evolutionary leap for mankind are perhaps most appropriately expressed in the following words of former UN Assistant Secretary General and Co-Founder of the United Nations University for Peace, Dr. Robert Muller:

**With Humanity's transcendence
into Cosmic Consciousness in Space and Time,
the Sciences and Religions will converge
into an All-Encompassing Holistic, Spiritual Age,
dreamt of
by our Ancestors, Prophets, Visionaries, Saints,
and the Earthly Incarnations
of Divine, Cosmic Forces over eons of time.**

~ Dr. Robert Muller, 2000—*Four Thousand Ideas for a Better World*

BASIC RESOURCES

Aerobus International—Dennis Stallings, President, 800 Rusk, Suite 1750, Houston, TX 77002-2814 (www.aerobus.com).

Doug Michels Studio—Doug Michels, World Architect, 3614 Montrose Blvd., Houston, TX 77006, (idealab@earthlink.net).

The GAIASHIP Project—Roar Bjerknes, Founder, President, R. B. Media (Pte. Ltd.), Moloveien 3A, 6004 Aalesund, NORWAY (www.gaiaship.org).

Integrity Research Institute—Tom Valone, Ph.D., ABD, President, 1220 L Street NW, Suite 100-232, Washington, DC 20005, (www.integrityresearchinstitute.org).

Planetary Association for Clean Energy—Andrew Michrowski, Ph.D., President, 100 Bronson Avenue, Suite 2001, Ottawa, Ontario K1R-6G8, CANADA (www.pacenet.homestead.com).

Rex Research—Robert A. Nelson, P. O. Box 19250, Jean, NV 89019.

SkvCat Technologies—3201 Airpark Drive, Suite 101, Santa Maria, CA 93455, (www.skycattech.com).

The Solar Tower Project—Roger Davey, Executive Director, Energen Global, Inc., 3717 E. Thousand Oaks Blvd., Westlake Village, CA 91362, (www.energenglobal.com).

The Venus Project—Jacque Fresco, Roxanne Meadows, 21 Valley Land, Venus, FL 33960, (www.thevenusproject.com).

ENDNOTES

INTRODUCTION

1. Michael Tobias, *World War II: Population and the Biosphere at the End of the Millennium,* Continuum Publishing Company, New York, 1998, p43.
2. David Suzuki, "Oceans Could Thrive Again," Environmental News Network, August 10, 2001; http://www.enn.com.
3. Thom Hartmann, *The Last Hours of Ancient Sunlight: Waking Up to Personal and Global Transformation,* Mythical Books, Northfield, VT, 1998, p.78.
4. Jeremy Lovell, "World Faces Critical Choices on Environment," Reuters, May 23, 2002; http://www.enn.com/news/wire-stories
5. *Winnipeg Free Press,* September 20, 2002; http://www.iids.org/task/force/documents.htm.
6. "Everest Glacier Melting, UN Climbers Find," Environment News Service, 6/11/02, http://www.ens-news.com.
7. "Kilimanjaro's White Peak May be a Memory by 2015," Environment News Service, 2/20/01, http://www.ens.lycos.com.
8. Environment News Service, June 11, 2002, http://www.ens.lycos.com.
9. Reuters News Service, March 20, 2002.
10. "Pollution is Plunging Us into Darkness," *New Scientist,* December 14, 2002, pp. 6-7, http://www.newscientist.com.
11. "Many of World's Lakes Face Death, Expert Warns," Reuters, November 12, 2002, http://www.enn.com/news/wire-stories.
12. "Northern European Cod Collapse Predicted," Environment News Service, October 29, 2002; http://www.ens-news.com.
13. "World's Seas Awash with Sewage," Environment News Service, October 5, 2002; http://www.ens-news.com.
14. "Pollution Kills Thousands of Children," Environment News Service, May 5, 2002; http://www.ens-news.com.

METASTRATEGY I
A NEW TRANSFORMATIVE EDUCATIONAL SYSTEM

15. Alvin Toffler, *Future Shock,* Random House, New York, 1970, pp. 342-343.

16. Bennett W. Goodspeed, *The Tao Jones Averages: A Guide to Whole-Brained Investing,* E. P. Dutton, Inc., New York, 983, p19.
17. John Gardner, *Self-Renewal: The Individual and the Innovative Society,* Harper and Row, New York, 1963, p. 25.
18. Leslie Walker, "In 18 Languages, Electronic Library Speaks to Children," *The Washington Post,* December 10, 2002; http://www.iht.com/articles/77800.html.
19. "Robot Aids in Trans-Atlantic Surgery," *USA Today,* September 20, 2001.
20. "Very Distant Learning," *The Futurist,* September-October, 2002, p.2.
21. Alfred Hermida, "Villagers Try Out Net on Wheels," BBC News, 7/28/02
22. "Computer Village in the Clouds," BBC News, October 22, 2001.
23. Yatri, *Unknown Man: The Mysterious Birth of a New Species,* Simon and Schuster, New York, 1988, p. 36.
24. Gerald O. Barney, *Global 2000 Revisited: What Shall We Do?* The Millennium Institute, Arlington, VA, 1993, p.5.

METASTRATEGY II
A GLOBAL RESOURCE MANAGEMENT PLAN

25. Ed Ayres, *God's Last Offer: Negotiating for a Sustainable Future,* Four Walls Eight Windows, New York, 1999, pp. 27-32.
26. Thom Hartmann, *The Last Hours of Ancient Sunlight: Waking Up to Personal and Global Transformation,* Mythical Books, Northfield, VT, 1988, p. 40.
27. "Marine Diseases: Symptoms of an Unhealthy Earth," Environmental News Network, May 30, 2001; http://www.enn.com.
28. Kieran Murray, "Mexico Needs 20 Years to Rescue Environment," Reuters, January 17, 2002; http://www.forests.org.
29. Reuters, "World Water Crisis Will Threaten One in Three," August 13, 2001.
30. "Satellite Aids in Monitoring Global Deforestation," Environment News Service, Ameriscan, June 3, 2001; http://www.ens.lycos.com.
31. Environment News Service, June 6, 2001, http://www.enn.com.
32. Mario Soares, Ed., *The Ocean Our Future: The Report of the Independent World Commission on the Oceans,* Cambridge University Press, Boston, 1998, p.79.
33. Environment News Service, March 13, 2001.

34. "Fires from Hell," *New Scientist,* August31, 2002, pp. 34-37; http://www.newscientist.com.
35. Robert Nelson, *Hemp Husbandry,* Morris Publishing, Kearney, NE, 1999, p.13.
36. Alan Bock, *The Orange County Register,* October 30, 1988.
37. John Roulac, *Industrial Hemp: Practical Products-Paper to Fabric to Cosmetics,* HEMPTECH, Ojai, CA, 1995, pp. 3-4.
38. Robert Nelson, "Bentonite Paper," Rex Research, Jean, NV, 1988.
39. Cliff Ransom, "Paper from Pond Scum", *Popular Science,* April, 2002, p. 30.
40. "Fastest-Growing Tree Not Catching On," Associated Press, July 10,2001;http://www.enn.com.
41. Charles C. Mann, "The Test Tube Forest," *Business 2.0,* February 2002; http://www.business2.com.
42. Lycos, December 7, 2000, "U.S. Plastic Lumber's Product is Used to Construct the World's First Suspension Bridge from Entirely Plastic Lumber;" http://www.ens.lycos.com.
43. Carlos Hernandez, and R. Mayur, *Pedagogy of Earth: Education for a Sustainable Future,* International Institute for Sustainable Future, Mumbai, India, 1999, pp. 34-35.
44. Elliott Norse and L. Watling, "Clearcutting the Ocean Floor," *Earth Island Journal,* Summer 1999, v.14, p. 2.
45. Dan Johnson, "The High Cost of Trawling," *The Futurist,* May 1999, p. 14.
46. Environment News Service, March 29, 2001; http://www.ens.com.
47. "Size Matters," *Rodale's Scuba Diving,* January-February, 2002; http://www.scubadiving.com.
48. "Other Size Benefits," Rodale's *Scuba Diving,* January-February, 2002.
49. "Coral Reefs Are an Oceanic Rarity," Environmental News Network, September 18, 2001;http://www.enn.com.
50. Ibid.
51. "Rigs to Reefs is the Heart of the Texas Artificial Reef Programs," Texas Parks and Wildlife, Austin Texas, June 30, 2002.
52. The Reef Ball Foundation; http://www.reefball.org.
53. Amanda Onion, "Reef Replacement: New Project Hopes to Restore Coral Reefs by Growing Them," abcnews.com, June 5, 2002.
54. The Geothermal Aquaculture Research Foundation; http://www.garf.org.
55. *New Scientist,* July 6, 2002, p. 38, http://www. newscientist.com.

56. *New Scientist,* September 29, 2001, p. 36;
    http://www.newscientist.com.
57. Caspar Henderson, "Electric Reefs," *New Scientist,* July 6, 2002,
    pp. 38-41, http://www.reefbase.org.
58. "Reef Fish Laundering Hides Pacific Overfishing," *Environment
    News Service,* July 15, 2002, http://www.ens-news.com.
59. Mario Soares, Ed., *The Ocean Our Future: The Report of the Inde-
    pendent World Commission on the Oceans,* Cambridge University
    Press, New York, 1998, p.124.

METASTRATEGY III
ENLIGHTENED POPULATION MANAGEMENT

60. Michael Tobias, *World War III: Population and the Biosphere at
    the End of the Millennium,* Continuum Publishing Company, New
    York, 1998, p.ll.
61. Thom Hartmann, *The Last Hours of Ancient Sunlight: Waking Up
    to Personal and Global Transformation,* Mythical Books, North-
    field, VT, 1998, p. 12.
62. "People Take up Most of the Planet, U.S. Study Says," Environmental
    News Network, October 10, 2002; http://www.enn.com.
63. Charlotte Denny, "Cows Are Better Off Than Half the World,"
    Earth, *The Guardian* [London], August 2002, p. 11.
64. Carlos Hernandez and R. Mayur, *The Pedagogy of Earth: Education
    for a Sustainable Future,* International Institute for Sustainable
    Future, Mumbai, India, 1999, pp. 32-33.
65. Ed Ayres, *God's Last Offer: Negotiating for a Sustainable Future,*
    Four Walls Eight Windows, New York, 1999, p. 78.
66. Paul H. Ray and S.R. Anderson, *The Cultural Creatives: How 50
    Million People Are Changing the World,* Harmony Books, New
    York, 2000, p. 215.
67. Alison Motluk, "Egg Stopper," *New Scientist,* November 30, 2000;
    http://www.newscientist.com.
68. "FDA OKs Nonsurgical Sterilization Device," Environmental
    News Network, November 10, 2002; htttp://www.enn.com.
69. http://www.cnn.com, 2000, http://www.seattlepi.com/local/96391_
    malepill20.shtml, 2002.
70. Michael Tobias, *World War III: Population and the Biosphere at
    the End of the Millennium,* Continuum Publishing Company, New
    York, 1998, p. 150.

71. Hannah Freeman, "Eco-Hotel Highlights the Best of Women's Work in Mexico," *E/Environmental Magazine,* February 1, 2002; http://www.enn.com/news.

72. "Women Found Eco Law in Iran," *Earth Island Journal,* Summer, 2002, p. 5.

73. Fred Pearce, "Lowering the Boom," *The Boston Globe,* December 17, 2002, http://www.boston.com

74. "Why We Need a Smaller U.S. Population, and How We Can Achieve It," Negative Population Growth, Inc. September 10, 2002, http://www.npg.org.

75. Carlos Hernandez and R. Mayur, *The Pedagogy of Earth: Education for a Sustainable Future,* International Institute for Sustainable Future, Mumbai, India, 1999, p. 83.

METASTRATEGY IV
A SHIFT TO CLEAN, RENEWABLE ENERGY TECHNOLOGIES

76. Sylvia Wright, 2000, "Trans-Pacific Pollution Worsens," August 1, MSNBC, 2000; http://www.msnbc.com/news.

77. "Mongolia Dust Cloud Moves across America," Environmental News Network, April 25, 2001; http://www.enn.com/news.

78. Ed Ayres, *God's Last Offer: Negotiating for a Sustainable Future,* Four Walls Eight Windows, New York, 1999, p. 17.

79. Mark Hertzgaard, *Earth Odyssey: Around the World in Search of Our Environmental* Future, 1998, p.169.

80. Cat Lazaroff, "Soot Called Major Cause of Global Warming," Environment News Service, February 8, 2001;

81. "Eliminating Soot Could Slow Global Warming," Reuters, December 12,2001;http://www.enn.com/news.

82. Russell Long, "Stacked Deck: Trading Air Quality for Global Trade," *Earth Island Journal,* Vol. 15 (4), Winter 2000-01.

83. "Soot, Particles, Aerosols: An Asian Brown Cocktail," Environment News Service, September 2, 2002; http://www.ens-news.com.

84. "L.A. Babies Get Lifetime's Toxic Air in Two Weeks," Environmental News Network, September 19, 2002; http://www.enn.com/news.

85. Robert Nelson, 1980, "Oil from Coal—Free: The Karrick LTC Process," Rex Research, Jean, NV.

86. Bob Sillery, "Tower of the Sun," *Popular Science,* February 20, 2003; http://www.popsci.com.

87. Michelle Nichols, "Australia Considers Giant Solar Power Tower," Reuters, January 7, 2003; http://www.enn.com/news.
88. Bill Moore, "Power Tube Taps Earth's Geothermal Resources," October 7, 2001; http://www.evworld.com.
89. "Geothermal Plant Could Power Entire UK," Environmental News Network, April 26, 2001; http://enn.com.
90. "JFK's 'Tidal Energy Dream,'" 3/5/03, Blue Energy Canada Inc.; http://www.bluenergy.com.
91. "Davis Hydro Turbine," March 5, 2003, Blue Energy, Canada Inc.;http://www.bluenergy.com/technology/turbione.html.
92. Sara Steindorf and Tom Regan, "New Turbine Can Extract Energy from Flowing Water," *Christian Science Monitor,* May 17, 2001, http://www.commondreams.org.
93. William McCall, "Electric Motor Efficiency Means Big Energy Savings," Associated Press, September 12, 2002; http://www.enn.com/news.
94. Jenny Hogan, "Now We Can Soak Up the Rainbow," *New Scien tist,* December 7, 2002; http://www.newscientist.com.
95. Mark Hertsgaard, *Earth Odyssey: Around the World in Search of Our Environmental Future,* Broadway Books, New York, 1998, p. 141.
96. Nuclear Solutions, Inc., 2002; http://www.nuclearsolutions.com.
97. Tathacus Resources, Ltd., Press Release, September 14, 2002. *Future Survey,* vol. 23, No. 12, December 2001, p. 9.
98. Andrew Michrowski, "Presentation by the Planetary Association for Clean Energy, Inc. Before the Select Committee on Alternative Fuels," Legislative Assembly of Ontario, Ottawa, January 30, 2002. *Earth Island Journal,* Summer, 2002, p.3.
99. "A Hydrogen Future," *Future Survey,* 23(12), December, 2001, p. 9. Christopher Doering, Reuters, Environmental News Network, June 25, 2002; http://www.enn.com/news.
100. Ruggero Maria Santilli, 2000—"Recycling Liquid Wastes and Crude Oil into MagneGas™ and MagneHydrogen™," Presented at the Hydrogen International Conference HY2000, Munich, Germany, September 11-15, 2000.
101. Dr. Bogdan Maglich, "Aneutronic Energy: A Search for Non-Radioactive Non-Proliferating Nuclear Power," The Tesla Foundation, P. O. Box 3037, Princeton, NJ 08543, November 3,2002; http://www.nuenergy.org.

102. Bill Morgan, "MEG Scalar Energy Device Patented—Production Starts Next Year," March 28, 2002. http://www.help4all.de/energy. MEGpsper.PDF.

METASTRATEGY V
A GLOBAL "GREEN REVOLUTION"

103. "A Climate Threshold Looms," *Earth Island Journal*, Summer, 2002, p. 3.
104. Christopher Doering, "Earth Can't Meet Human Demand for Resources, Says Study," Reuters, June 25, 2002; http://enn.com/news.
105. Mark Hertsgaard, "A Global Green Deal," *Time*, April-May, 2000.
106. J. R. Pegg, "Donating More than Money," Environment News Service, November 5, 2002; http://ens-news.com.
107. "These Roofs Last a Lawn Time," *Earth Island Journal*, VI7 (2), Summer, 2002, p. 3.
108. MYCOVA, "Helping the Ecosystem through Mushroom Cultivation;" http://www.fungi.com.
109. "Australia's Catholic Church Embraces Environment," Environment News Service, July 2, 2002; http://ens-news.com.
110. "Website Calculates Trees Felled for Magazine Printing," Environment News Service, August 30, 2002; http://ens-news.com.
111. "Clean up the World," *Earth Island Journal*, Winter 2002-3, p. 10.
112. "Peru Preserves Untouched Andean Rainforest in New National Park," Environment News Service, May 25, 2001; http://ens-news.com.
113. "Peru Swaps Debt for Tropical Rainforest Protection," Environment News Service, June 26, 2002; http://ens-news
114. Michael Astor, Associated Press, "Brazil Creates World's Largest Tropical National Park," August 23, 2002; http://www.enn.com.
115. "Australia Creates World's Largest Marine Reserve," Environment News Service, October 9, 2002; http://ens-news.com.
116. David Ljunggren, Reuters, "Canada to Create 10 Enormous New National Parks," October 4, 2002; http://www.enn.com/news.
117. Ariadna Reida, G. Cook, "Baikal Watch: Siberian Parks in Peril," *Earth Island Journal*, Winter 2002-3, p. 19.

METASRTATEGYVI
NEW GLOBAL STRATEGIES FOR WASTE TREATMENT

118. John Robbins, *Diet for a New America: How Your Food Choices Affect Your Health, Happiness, and the Future of Life on Earth,* H. J. Kramer, Inc., Tiburon, CA, 1987, p. 372.

119. International Bio-Recovery Corporation, North Vancouver, British Columbia, Canada; http://www.ibrcorp.com.

120. Andy Coughlan, "The House that Cack Built," *New Scientist,* August 31, 2002, p. 18; http://www.newscientist.com.

121. Vicki Smith, "Chicken Manure Could Become Environmentally Friendly Fuel,"Assocjated Press, December 18, 2001; http://enn.com/news/wire-stories.

122. "Pachyderm Paper Makes a Statement," BBC News, South Asia, July 5, 2001; http://news.bbc.co.uk/lows/english/world/southasia.

123. Gordon Solberg, *Building with Papercrete and Adobe: A Revo lutionary New Way to Build Your Own Home for Next to Nothing,* Remedial Planet Communications, Radium Springs, NM, 1999.

124. Jason G. Daley, "Green Burial at Sea," *E-Magazine,* January-February, 2002, p. 13; http://www.eternalreefs.com.

125. "UK Looks to Bacteria for Pollution Solutions," Environment News Service, April 6, 2001; http://www.ens-news.com.

126. "Hydro-Environmental Resources Announces Commercial Availability of Low-Pressure Reactor," October 21, 2002; http://www.hydrogenerate.com.

127. "From Sewage to Wattage," *Earth Island Journal,* Spring, 2002, p. 42.

128. "Japan Tops Can Returns," *Living Planet,* Winter 2001, p. 9.

129. Daniel Imhoff, "Thinking outside the Box: A Systems View of Packaging," *Whole Earth,* Winter 2002, p. 9.

130. Jim Motavalli, "Zero Waste," *E-Magazine,* March-April, 2001, p. 31.

131. Paul Hawkin, A. Lovins, A. H. Lovins, *Natural Capitalism: Creating the Next Industrial Revolution,* Little Brown & Co., New York, 1999, pp. 52-53.

132. Daniel Imhoff, op. cit., p. 12.

133. Daniel Imhoff, op. cit., pp. 16-17.

134. David Kupfer, *Whole Earth,* "Have Your Plate and Eat it Too," Winter 2002, p. 17.

135. Ibid.; http://www.biotrem.com.pl.en//.

136. Ibid.

137. http://www.potatoplates.com.
138. http://www.novamont.com.
139. http://www.edibowls.com.
140. http://www.earthshell.com.
141. David Kupfer, op. cit.
142. *Living Planet,* Winter, 2001, p. 9.
143. "First Compostable Logo Goes to Biocorp Plastic Bags," Environmental News Network, June 12, 2001; http://www.enn.com.
144. "Biodegradable Plastics Could Reduce Landfill Need," Environment News Network, September 10, 2002; http://ens-news.com.
145. "Recycled Plastics Form Bridge Beams," Environment News Service, January 20, 2003; http://ens-news.com.
146. Colin Woodward, "Fair Winds in Denmark: A Scandinavian Pioneer Shows that Sustainability Can be Profitable," *E, Environmental Magazine,* July-August, 2001.

METASTRATEGY VII
SUSTAINABLE HUMAN ECOLOGIES

147. John L. Peterson, *The Road to 2015: Profiles of the Future,* Waite Group Press, 1994, p. 170.
148. John N. Ott, *Health andLight: The Effects of Natural and Artificial Light on Man and Other Living Things,* Pocket Books, New York, 1973.
149. James Carey and M. Carey, "Indoor Pollution Can Affect Health," Associated Press, February 19, 2002; Http://enn.com.
150. Joel Schwartz, "World Gets Bleaker for our Children, Contends Psychologist," January 27, 2003; http://www.washington.edu/newsroom.
151. "Trees Save San Antonio Millions Each Year," Environment News Service, November 13, 2002; http://ens-news.com.
152. "Carfree Cities," September 20, 2002, http://www.carfree.com.
153. "Farewell Fast Lane: Italy Incorporates 'Slow Cities;'" http://www.cnn.com/2000/WORLD/europe/italy.slowcities.ap/.
154. "World's Largest Man Made Islands to be Created in Dubai," *Coastweek;* http://www.coastweek.com.
155. Hope Cristol, "A Maritime Utopia," *The Futurist,* November-December, 2002, p. 68-69.
156. Courtney McCall, *Texas Technology,* June, 2001, pp. 60-61.
157. "From Telecommuting to Teleporting," *The Futurist,* May-June, 2001, pp. 34-35.

158. Jacque Fresco, *The Best That Money Can't Buy: Beyond Politics, Poverty, & War,* Global Cybervisions, Venus, FL, 2002, p. x.
159. Jacque Fresco, *The Venus Project: The Redesign of a Culture,* Global Cyber-Visions, Venus, FL, 1995.
160. Jacque Fresco, 1995, op. cit., p. 5.
161. Jacque Fresco and R. Meadows, *Cities in the Sea* (Video Presentation), Global Cyber-Visions, Venus, FL, 2002; http://www.thevenusproject.com.
162. Jacque Fresco, 1995, op. cit., p. 8.
163. Jacque Fresco, 2002, op. cit., p. 123.
164. Jacque Fresco, ibid, p. 103.

METASTRATEGY VIII
CLEAN, ENERGY-EFFICIENT TRANSPORTATIONS SYSTEMS

165. Rocky Mountain Institute, "The Hypercar Concept;" http://www.rmi.org.
166. Carolyn Dempster, "The Car That Runs on Air," BBC News, October 24, 2000; http://us.f404.mail.yahoo.com.
167. Jim Starry, "A 'Green Airport'? Meet the StarPort," Earth Island Journal, Autumn, 2001,V16 (3), p. 38; http://www.earthisland.org.
168. SkyCat Technologies, Inc., 3201 Airpark Drive, Suite 101, Santa Maria, CA 93455; http://www.skycattech.com.
169. Lynn Cook, "Oh Heavens!" *Forbes,* December 11, 2000, pp. 204-5, Aerobus International, Inc.; http://www.aerobus.com.
170. Russell Long, "Stacked Deck: TradingAir Quality for Global Trade," *Earth Island Journal,* 15(4): Winter, 2000-01.
171. Lyndon LaRouche, *The LaRouche Program to Save the Nation,* LaRouche's Committee for a New Bretton Woods, Leesburg, VA, 1998, pp. 26-32.

METASTRATEGY IX
A SHIFT FROM CONSUMPTION-DRIVEN TO ECO-APPROPRIATE LIFESTYLES

172. Ed Ayres, *God's Last Offer: Negotiating for a Sustainable Future,* Four Walls Eight Windows, New York,1999, p. 27
173. Ibid., p. 36.
174. Elizabeth Rosenthal, "The Rage in China: Lunching on Wildlife and Mao," *New York Times International,* June 25, 2002.
175. Michelle McLean and C. Woods, "Aquaculture and Fisheries Enhancement: Seahorses Culture," *Aquaculture Update,* No. 20, Summer 1998; http://www.niwa.cri.nz/articles/fishieries2.html.

176. Paul H. Ray and S. R. Anderson, *The Cultural Creatives: How 50 Million People Are Changing the World,* Harmony Books, New York, 2000.

177. Ibid., pp. 166-167.

178. Emil Polk, "Ultraviolet for Health," *Whole Earth,* Winter 2000, p. 39; http://www.waterhealth.com.

179. Emily Lambert, "Running Dry," *Forbes,* January 6, 2003, pp. 144-145.

180. "Shantytown Recycles for Profit," *Earth Island Journal,* Summer 2002, p. 5.

181. Christian Kuchii, *Forests of Hope: Stories of Regeneration,* New Society Publishers, Gabriola Island, B.C., Canada, 1997, p. 34.

METASTRATEGY X
WORLDWIDE SHIFT FROM MEAT- TO PROTEIN-BASED DIETS

182. Jim Mason and P. Singer, *Animal Factories,* Crown Publishers, New York, 1980, p. 84.

183. Raymond Loehr, *Pollution Implications of Animal Wastes—A Forward Oriented Review,* Water Pollution Control Research Series, Washington, DC, Office of Research and Monitoring, U.S.E.P.A., 1968, p. 26.

184. John Robbins, *Diet for a New America: How Your Food Choices Affect Your Health, Happiness, and the Future of Life on Earth.* H. J. Kramer, Tiburon, CA, 1987, p. 372.

185. Christopher Finn, "The Real Cost of Shrimp," United Nations Food and Agriculture Organization; Natural Resources Defense Council, National Marine Fisheries Service, Date Unknown.

186. "It's a Fish-Eat-Fish World," *Earth Island Journal,* Winter 2002, p. 14.

187. Mark J. Palmer, "'Dolphin-Safe' Label Victory," *Earth Island Journal, Winter 2001-2.*

188. *USA Today,* June 4, 2001, *Utne Reader,* Sept.-Oct., 1999.

METASTRATEGY XI
STRATEGIES FOR CREATION AND DISTRIBUTION OF SUPERFOODS

189. Cat Lazaroff, "Food Travels Far to Reach Your Table," Environment News Service, 11/23/02; http://ens-news.com; http://www.worldwatch.org.

190. Vithoon Amorn, "Insects Invade Thai Supermarkets," Reuters, May 28, 2002; Environmental News Network; http://www.enn.com.

191. Christopher Hills and H. Nakamura, *Food from Sunlight: Planetary Survival for Hungry People,* University of the Trees Press, Boulder Creek, CA, 1978.
192. "Designer Diets Produce Fish Packed with Nutrients," Environment News Service, January 8, 2003; http://ens-news.com.
193. Michael Zey, *Seizing the Future: The Dawn of the Macroindustrial Era,* 1998, pp. 169-171.
194. Wendy Wolfson, "Raising the Steaks," *New Scientist,* December 21-28, 2002, p. 60; http://www.newscientist.com.
195. Michael Zey, 1998, op. cit., p. 171.
196. Wendy Wolfson, op. cit., p. 62.
197. Australian Broadcasting Corporation, January 20, 1999; http://www.abc.net.

METASTRATEGY XII
NEW APPLICATIONS OF CONSCIOUSNESS TECHNOLOGY

198. Edward Russell, *Report on Radionics: Science of the Future,* C. W. Daniel Company, Essex, UK, 1973.
199. David V. Tansley, *Radionics Interface with the Ether Fields,* C. W. Daniel Company, Essex, UK, 1975.
200. Environment News Service, "Environmental Treaties Enforced from the Heavens," Environment News Service, April 25, 2001; http://ens.lycos.com/ens.
201. Stephen M. Phillips, *Extra-Sensory Perception of Quarks,* Theosophical Publishing House, Wheaton, IL, 1980.
202. Larry Dossey, "Measuring the Power of Prayer," *Whole Earth,* Spring 2002, pp. 20-21.
203. Constantin I. Ivanenko, Founder of Russian Initiative for Defense of Earth, St. Petersburg, Russia, Personal Communication, 2001-2003.
204. P. M. H. Atwater, *Future Memory: How Those Who "See the Future" Shed New Light on the Workings of the Human Mind,* Birch Lane Press, New York, 1996.

THE FUTURE OF THE GLOBAL BIOSPHERE:
PERSPECTIVES AND POSSIBILITIES

205. "Australia's Great Barrier Reef Recovers from Disease to Become One of the World's Healthiest," Environmental News Network, December 18, 2002; http://www.enn.com.

206. "Africa's Hidden Waters," *The Futurist,* September-October, 2002, p. 2; http://www.unesco.org.

207. "Something Fishy," *Earth Island Journal,* Summer 2002, p. 2.

208. "Elephant Band Raises Awareness of Wildlife Plight," Environmental News Network, April 23, 2002; http://www.enn.com.

209. "Swordfish Stage a Strong Recovery," Environment News Service, October 8, 2002; http://ens-news.com.

210. Michael Perry, "Antarctic Ozone Hole Could Close by 2050, Scientists Reports," Reuters, Environmental News Network, September 18, 2002; http://enn.com.

211. "World's Oldest Trees to be Cloned," Environment News Service, October 5, 2002; http://ens-news.com.

212. Robert Muller, *Four Thousand Ideas for a Better World,* Global Public Relations, Santa Barbara, CA, 2000; http://www.worldpeace.org/ideas/.

BIBLIOGRAPHY

Abdullah, Sharif, 1999—**Creating a World that Works for All.** Berrett-Koehler Publishers, Inc., San Francisco.

Alvord, Katie, 2000—**Divorce Your Car: Ending the Love Affair With the Automobile.** New Society Publishers, Gabriola Island, B.C., Canada.

Attfield, Robin, 1999—**The Ethics of the Global Environment.** Purdue University Press, West Lafayette, IN.

Atwater, P. M. H., 1996—**Future Memory: How Those Who "See The Future" Shed New Light on the Workings of the Human Mind.** Hampton Roads Publishing Company, Inc. Charlottesville, VA.

Avery, Samuwl, 1995—**The Dimensional Structure of Consciousness: A Physical Basis for Immaterialism.** Compari, Lexington, KY.

Ayres, Ed, 1999—**God's Last Offer: Negotiating for a Sustainable Future.** Four Walls Eight Windows, New York.

Baker, Robin, 1999—**Sex in the Future: The Reproductive Revolution and How It Will Change Us.** Arcade Publishing, Inc., New York.

Barney, Gerald O., 1993—**Global 2000 Revisited: What Shall We Do? The Critical Issues of the 21st Century.** Millennium Institute, Arlington, Virginia.

Barney, Gerald O., 1999—**Threshold 2000: Critical Issues and Spiritual Values for a Global** Age. Millennium Institute, Arlington, VA.

Bates, Albert K., 1990—**Climate in Crisis: The Greenhouse Effect And What We Can Do.** The Book Publishing Company, Summertown, TN.

Bennett, Frederick, 1999—**Computers as Tutors: Solving the Crisis in Education.** Faben, Inc., Sarasota, FL.

Berger, John J., 2000—**Beating the Heat: Why and How We Must Combat Global Warming.** Berkeley Hills Books, Berkeley, CA.

Berrill, Michael, 1997—**The Plundered Seas: Can the World's Fish Be Saved?** Greystone Books, Vancouver, BC, Canada.

Brown, Lester R., 2001—**Eco-Economy: Building an Economy for the Earth.** W. W. Norton and Company, New York.

Collinge, William, 1998—**Subtle Energy: Awakening the Unseen Forces in Our Lives.** Warner Books, New York.

Dauncey, Guy, 2001—**Stormy Weather: 101 Solutions to Global Climate Change.** New Society Publishers, Gabriola Island, B.C., Canada.

Davidson, Osha Gray, 1998—**The Enchanted Braid: Coming to Terms with Nature on the Coral Reef.** John Wiley & Sons, Inc., New York.

De Graaf, John, D. Wann, and T. H. Naylor, 2001—**Affluenza: The All-Consuming Epidemic.** Berrett-Koehler Publishers, Inc., San Francisco, CA.

Devreaux, Paul, 1992—**Secrets of Ancient and Sacred Places.** Sterling Publishing Company, New York, NY.

Devreaux, Paul, J. Steele and D. Kubrin, 1989—**Earthmind: Communicating with the Living World of Gaia.** Destiny Books, Rochester, VT.

Fresco, Jacque, 1995—**The Venus Project: The Redesign of a Culture.** Global Cyber-Visions, Venus, FL.

Fresco, Jacque, 2002—**The Best That Money Can't Buy: Beyond Politics, Poverty, and War.** Global Cyber-Visions, Venus, FL.

Gardner, John W., 1963—**Self-Renewal: The Individual and the Innovative Society.** Harper and Row, New York.

Glenn, Jerome C. and T. J. Gordon, **2002—2002 State of the Future.** American Council for the United Nations University, Washington, DC.

Hagelin, John, 1998—**Manual for a Perfect Government: How to Harness the Laws of Nature to Bring Maximum Success to Governmental Administration.** Maharishi University of Management Press, Fairfield, IA.

Hawkin, Paul, 1993—**The Ecology of Commerce: A Declaration of Sustainability.** HarperCollins Publishers, New York.

Hawkin, Paul, A. Lovins, A. H. Lovins, 1999—**Natural Capitalism: Creating the Next Industrial Revolution.** Little Brown & Company, New York

Hernandez, Carlos and R. Mayur, 1999—**Pedagogy of the Earth: Education for a Sustainable Future.** International Institute for Sustainable Future, Mumbai, India.

Hartmann, Thom, 1998—**The Last Hours of Ancient Sunlight.** Mythical Books, Northfield, VT.

Hertsgaard, Mark, 1998—**Earth Odyssey: Around the World in Search of Our Environmental Future.** Broadway Books, New York.

Hills, Christopher, and H. Nakamura, 1978—**Food from Sunlight: Planetary Survival for Hungry People.** University of the Trees Press, Boulder Creek, CA.

Honey, Martha, 1999—**Ecotourism and Sustainable Development: Who Owns Paradise?** Island Press, Washington, DC.

Houston, Jean, 2000—**Jump Time: Shaping Your Future in a World of Radical Change.** Jeremy P. Tarcher/Putnam, New York.

Hunt, Valerie V., **1989—Infinite Mind: The Science of Human Vibrations.** Malibu Publishing Company, Malibu, CA.

Katra, Russell and R. Targ, 1998—**Miracles of the Mind: Exploring Nonlocal Consciousness and Spiritual Healing.** New World Library, Novato, CA.

Kay, Jane Holtz, 1997—**Asphalt Nation: How the Automobile Took Over America and How We Can Take It Back.** University of California Press, Berkeley, CA.

King, Serge K., 1992—**Earth Energies: A Quest for the Hidden Power of the Planet.** Quest Books, Wheaton, IL.

Kuchii, Christian, 1997—**Forests of Hope: Stories of Regeneration.** New Society Publishers, Gabriola Island, B.C., Canada.

LaRouche, Lyndon, 1998—**The LaRouche Program to Save the Nation.** LaRouche's Committee for a New Bretton Woods, Leesburg, VA.

Laszlo, Ervin, 2001—**Macroshift: Navigating the Transformation to a Sustainable World.** Berrett-Koehler Publishers, Inc., San Francisco.

Levy, Pierre, 1997—**Collective Intelligence: Mankind's Emerging World in Cyberspace,** Perseus Books, Cambridge, MA.

Lovelock, James, 1991—**Gaia: The Practical Science of Planetary Medicine.** Oxford University Press.

Little, Charles E., 1995—**The Dying of the Trees: The Pandemic in America's Forests.** Penguin Books, New York.

Mason, Jim and P. Singer, 1980—**Animal Factories.** Crown Publishers, New York.

Metzner, Ralph, 1999—**Green Psychology: Transforming Our Relationship to the Earth.** Park Street Press, Rochester, VT.

McMoneagle, Joseph, 1998—**The Ultimate Time Machine: A Remote Viewer's Perception of Time and Predictions for the New Millennium.** Hampton Roads Publishing Company, Charlottesville, Virginia.

Milani, Brian, 2000—**Designing the Green Economy: The Postindustrial Alternative to Corporate Globalization.** Rowman and Littlefield Publishers, New York.

Muller, Robert, 2000—**Four Thousand Ideas for a Better World.** Global Public Relations, Santa Barbara, CA.

Nelson, Robert, 1999—**Hemp Husbandry.** Morris Publishing, Kearney, NE.

Osborne, Michael J., 2001—**Lightland: Climate Change and the Human Potential.** Plain View Press, Austin, TX.

Ott, John N., 1973—**Health and Light: The Effects of Natural and Artificial Light on Man and Other Living Things.** Simon and Schuster, New York.

Peterson, John L., 1994—**The Road to 2015: Profiles of the Future.** Waite Group Press, Courte Madera, CA.

Pettis, Chuck, 1999—**Secrets of Sacred Space.** Llewellyn Publications, St. Paul, MN.

Phillips, Stephen, 1980—**Extra-Sensory Perception of Quarks.** The Theosophical Publishing House, Wheaton, IL.

Pogacnik, Marko, 1997—**Healing the Heart of the Earth: Restoring the Subtle Levels of Life.** Findhorn Press, Tallahassee, FL.

Porritt, Jonathon, 2000—**Playing Safe: Science and the Environment.** Thames and Hudson, New York, p. 87.

Ray, Paul H. and S. R. Anderson, 2000—**The Cultural Creatives: How 50 Million People Are Changing the World.** Harmony Books, New York, NY.

Radin, Dean, 1997—**The Conscious Universe: The Scientific Truth of Psychic Phenomena.** HarperCollins Publishers, Inc., New York.

Rifat, Tim, 2001—**Remote Viewing: What It Is, Who Uses It and How to Do It.** Vision Paperbacks, London.

Robbins, John, 1987—**Diet for a New America: How Your Food Choices Affect Your Health, Happiness, and the Future of Life on Earth.** H. J. Kramer, Inc., Tiburon, CA.

Roseland, Mark, Ed., 1997—**Eco-City: Healthy Communities, Healthy Planet.** New Society Publishers, Gabriola Island, British Columbia, Canada.

Roulac, John, 1995—**Industrial Hemp: Practical Products— Paper to Fabric to Cosmetics.** HEMPTECH, Ojai, CA.

Russell, Edward W., 1973—**Report on Radionics: Science of the Future.** C. W. Daniel Company, Ltd., Essex, United Kingdom.

Russell, Peter, 1995—**The Global Brain Awakens: Our Next Evolutionary Leap.** Global Brain, Inc., Palo Alto, CA.

Schneider, Kenneth R., 1971—**Autokind vs. Mankind,** W. W. Norton and Company, New York.

Shiva, Vandana, 2002—**Water Wars: Privatization, Pollution, and Profit.** South End Press, Cambridge, MA, 2002.

Smil, Vaclav, 1999—**Energies: An Illustrated Guide to the Biosphere and Civilization.** MIT Press, Cambridge, MA.

Soares, Mario, Ed., 1998—**The Ocean Our Future: The Report of the Independent World Commission on the Oceans.** Cambridge University Press, Boston.

Solberg, Gordon, 1999—**Building with Papercrete and Paper Adobe: A Revolutionary New Way to Build Your Own**

Home for Next to Nothing. Remedial Planet Communications, Radium Springs, NM.

Suzuki, David, 1989—**Inventing the Future: Reflections on Science, Technology and Nature.** Stoddart Publishing Company, Ltd., Toronto, Canada.

Tansley, David V., 1975—**Radionics Interface with Ether Fields.** The C. W. Daniel Company, Ltd., Essex, England.

Targ, Russell and J. Katra, 1998—**Miracles of the Mind: Exploring Nonlocal Consciousness and Spiritual Healing.** Hampton Roads Publishing Company, Inc., Charlottesville, VA.

Tart, Charles T., Ed., 1997—**Body, Mind, Spirit: Exploring the Parapsychology of Spirit.** Hampton Roads Publishing Company, Inc. Charlottesville, VA.

Tobias, Michael, 1998—**World War III: Population and the Biosphere at the End of the Millennium.** Continuum Publishing Company, New York.

Toffler, Alvin, 1970—**Future Shock.** Random House, New York.

Van Driesche, Jason and R. Van Driesche, 2000—**Nature Out of Place: Biological Invasions in the Global Age.** Island Press, Washington, DC.

Wann, David, 1990—**Bio-Logic: Designing With Nature to Protect the Environment.** Johnson Books, Boulder, CO.

Watson, Lyall, 1988—**Beyond Supernature: A New Natural History of the Supernatural.** Bantam Books, New York.

White, John and S. Krippner, Eds., 1977—**Future Science: Life Energies and the Physics of Paranormal Phenomena.** Anchor Books, Garden City, New York.

Yatri, 1988—**Unknown Man: The Mysterious Birth of a New Species.** Simon and Schuster, New York.

Zey, Michael G., 1994—**Seizing the Future: How the Coming Revolution in Science, Technology, and Industry Will Expand the Frontiers of Human Potential and Reshape the Planet.** Simon and Schuster, New York.

Zohar, Danah, 1990—**The Quantum Self: Human Nature and Colnsciousness Defined by the New Physics.** William Morrow, New York.

SECOND EDITION REFERENCES

Begich, Nick, 2006—HAARP-The Update: Angels Still Don't Play this HAARP (DVD). Earthpulse Press, Inc., Anchorage, AK (www.earthpulse.com).

Brohy, Audrey and G. Ungerman, 2008—Belonging: A Scientific and Spiritual Journey into Humanity's Footprint on the Earth (DVD). Free-Will Productions USA, P. O. Box 3055, Chico, CA 95027 (www.freewillprods.com, www.belongingthedoc.com).

Gore, Albert, 2007—An Inconvenient Truth: A Global Warning (DVD). Paramount Pictures, Hollywood, CA. (www.climatecrisis.net).

Laszlo, Ervin, 2006—The Chaos Point: The World at the Crossroads. Hampton Roads Publishing Co, Inc., Charlottesville, VA.

Laszlo, Ervin, 2006—Science and the Reenchantment of the Cosmos: The Rise of the Integral Vision of Reality. Inner Traditions, Rochester, VT.

Laszlo, Ervin, 2008—Quantum Shift in the Global Brain: How the New Scientific Reality Can Change Us and Our World. Inner Traditions. Rochester, VT.

Laszlo, Ervin and J. Currivan, 2008—Cosmos: A C-Creator's Guide to the Whole World. Hay House, Inc., Carlsbad, CA.

Mannning, Jean and J. Garbon, 2009—Breakthrough Power: Quantum-Leap New Energy Inventions. Amber Bridge Books, Vancouver, BC, Canada.

O'Leary, Brian, 2008—The Energy Solution Revolution. Montesuenos Press, P. O. Box 258, Loja, Ecuador (www.brianoleary.com).

Valone, Thomas, 2008—The Future of Energy: An Emerging Science. Integrity Research Institute, 5020 Sunnyside Ave., Suite 209, Beltsville, MD 20705 (www.IntegrityResearch.org).

Valone, Thomas, 2009—Zero Point Energy, the Fuel of the Future. Integrity Research Institute, 5020 Sunnyside Ave., Suite 209, Beltsville, MD 20705 (www.IntegrityResearch.org).

Elliott Maynard:
Biographical Overview

Elliott's educational background includes a Bachelor of Arts Degree from Washington and Lee University, a Master of Science degree in Zoology from the University of Miami, and advanced graduate studies in Coral Reef Ecology, Biological Oceanography, Phytoplankton Ecology, and Tropical Rainforest Ecology at the University of Miami Marine Institute, Nova Oceanographic Center, and the University of Costa Rica. Elliott received a Ph.D. Degree in Consciousness Research at the University of the Trees in Boulder Creek, California, and has also served as a faculty member at Adelphi University and Dowling College in New York.

Elliott's scientific career includes intensive field research on tropical marine ecosystems and on the behavior of coral reef shrimps and fishes. He studied Tropical Biology and Rainforest Ecology at the Organization for Tropical Studies in Costa Rica with renowned global forestry expert, Dr. Leslie Holdridge, and continued on to do pioneering research and development of giant freshwater shrimp culture in the Florida Everglades and Puerto Rico. In 1980-81 he worked as a Research Scientist at the Kuwait Institute for Scientific Research in conjunction with 1200 scientists and technical specialists from around the world. In Kuwait he played key roles in Project Management, Development of Advanced Shrimp Culture Systems, and Future Food Resources.

In 1978, Elliott and his wife, Sharon founded Arcos Cielos Research Center in Sedona, Arizona. This international non-profit center for Science, Education, Fine Arts, Global Ecology, and Human Potential Development continues to be actively engaged in developing unique new paradigms in Education, Fine Arts,

Global Ecology, Alternative Energy, Transformative Thinking, and Global Conflict Resolution.

Elliott's contributions to education include a position as Consultant Coordinator and Secondary Education Specialist for the Hopi Tribe in Northern Arizona. In 1978 he assisted in developing a bi-cultural education programs for the Hopi Elementary School System, and a new $22 million Jr. /Sr. High School. For over 25 years Elliott has been active in the field of Management Consulting, working with corporate leaders, celebrities, and key professionals to develop new programs for heightened efficiency and breakthrough thinking. He holds the titles of APC (Accredited Professional Consultant), CC (Certified Consultant), CPC (Certified Professional Consultant), and CPCM (Certified Professional Consultant to Management).

Elliott's experience also includes the integration of alternative medical and healing technologies with conventional medicine. His recent efforts have focused on developing effective solutions for the major problems of our global ecosystems (Planetary Management), through practical integration of alternative technologies with conventional science. In addition to being an accomplished underwater photographer, videographer and scuba diver, Elliott has done extensive observational research in the marine environment. From 2000 to 2005, he visited remote locations in the South Pacific Islands and atolls to gain first-hand knowledge of the current health of the marine and coral-reef ecosystems. He also made a series of dives to film shark feeding behavior in French Polynesia.

Elliott is a member of the Aerospace Technology Working Group and has been an active participant in their teleconferences, workshops, and presentations. ATWG consists of an elite group of scientists and key leaders in NASA, academia, the commercial aerospace community, scientific research institutions, and the US Military. This organization functions as a unique think-tank and strategic advisory resource for developing new applications of space science, to enrich our knowledge of earth-based environmental problems and to develop advanced technologies for expanding the human presence in orbital and outer space. Elliott also serves on the Advisory Board of the Washington, DC based Integrity Research Institute, an organization which focuses on developing advanced technologies in areas of alternative energy, global ecology, and advanced medical techniques.

More recently, Elliott has been working with the London-based organization, Humanitad International, to develop and coordinate new strategies for implementing advanced non-polluting technologies, to develop creative programs for implementing world peace, and to encourage international and intercultural cooperation. In this regard, he functions as Director of Advanced Technology and Global Ecology for this organization, which works in conjunction with the United Nations Millennium Development Goals Awards Events, and also in areas of Global Conflict Resolution and International Cultural Exchange. Besides being founder and president of Arcos Cielos Corporation, a non-profit 501(c)(3) corporation, he is also president of Elshar Executive Enterprises, Inc. and Elshar International, LLC, which were established for individual and corporate consulting.

Elliott continues working to develop unique new paradigms for thought and action, which are an extension of his 2003 book, *Transforming the Global Biosphere: Twelve Futuristic Strategies*. These new paradigms include: a new environmentally focused educational system, The Arcos Cielos University of the Future Project® and Future-Science Art®: The Creation of Living Artworks for the Third Millennium®. He is currently writing a new book called *Working in the Quantum Field: For Personal and Global Transformation*, and is also in the process of developing a new Master Paradigm within the context of quantum-field thinking which he calls Future-Science Technology®.

Breinigsville, PA USA
01 November 2009
226820BV00001B/1/P